BOOKS BY RITCHIE WARD

Into the Ocean World
(*1974*)

The Living Clocks
(*1971*)

Practical Technical Writing
(*1968*)

THESE ARE BORZOI BOOKS,
PUBLISHED IN NEW YORK BY
ALFRED A. KNOPF

INTO
THE
OCEAN
WORLD

Ritchie Ward

INTO THE OCEAN WORLD

The Biology of the Sea

Alfred A. Knopf

New York · 1974

THIS IS A BORZOI BOOK
PUBLISHED BY ALFRED A. KNOPF, INC.

First Edition

Copyright © 1974 by Ritchie R. Ward

All rights reserved under International and Pan-American Copyright Conventions. Published in the United States by Alfred A. Knopf, Inc., New York, and simultaneously in Canada by Random House of Canada Limited, Toronto. Distributed by Random House, Inc., New York.

Library of Congress Cataloging in Publication Data

Ward, Ritchie R.
Into the ocean world.

Bibliography: p.
1. Marine biology. 2. Oceanography. I. Title.
QH91.W35   574.92   73–18033
ISBN 0-394-47405-8

Manufactured in the United States of America

To
Bill and Anne,
Billy, Heidi, Shannon,
Coleen, Katie,
Arrow, and George

Man will always pursue; he will always find; and he will always be different tomorrow from what he is today.
MELVIN CALVIN, 1969

CONTENTS

ILLUSTRATIONS   *xi*

PREFACE   *xv*

1. MAN, SCIENCE, AND THE SEA   *3*
2. HAWAII-LOA: WITHOUT COMPASS OR SEXTANT   *11*
3. COOK: AROUND THE WIDE PACIFIC   *27*
4. PRIESTLEY AND INGENHOUSZ: "THE KEY CHEMICAL REACTION OF THE UNIVERSE"   *43*
5. FORBES: THE DAWN OF MARINE ECOLOGY   *57*
6. LOUIS AGASSIZ: "AN INTERCOURSE WITH THE HIGHEST MIND"   *71*
7. MAYER: THE FORCES OF LIFE   *85*
8. THOMSON: THE DEPTHS OF THE SEA   *95*

9. MURRAY: PIONEER OF OCEANOGRAPHY   *113*

10. THE LABORATORY COMES TO THE SEA   *137*

11. ALEXANDER AGASSIZ: COPPER MINES AND CORAL REEFS   *152*

12. HERDMAN: OCEANOGRAPHY IN THE SERVICE OF MAN   *168*

13. HJORT AND HENSEN: THE DYNAMICS OF OCEAN POPULATIONS   *185*

14. RICKETTS: THE ECOLOGY OF THE LITTORAL   *195*

15. HARDY: FOOD CHAINS OF THE SEA   *212*

16. ZOBELL: MICROBES OF THE SEA   *230*

17. CALVIN: NEW INSIGHTS INTO PHOTOSYNTHESIS   *241*

18. ON SEEKING A POINT OF VIEW   *256*

19. BENSON: ENERGY FLOW IN OCEAN LIFE   *266*

20. CONSERVING THE BOUNTY OF THE SEA   *280*

21. "THE EFFECTING OF ALL THINGS POSSIBLE"   *302*

SOURCES   *310*

INDEX   *to follow page*   *323*

ILLUSTRATIONS

A lone clam-digger. Photograph by the author   *5*
Dredging marine samples from the shallow bottom. Courtesy *Oakland (California) Tribune*   *8*
The Tuamotu Archipelago in the South Pacific. Courtesy National Aeronautics and Space Administration   *13*
Bow of an ancient Tuamotuan sailing canoe. Courtesy Bernice P. Bishop Museum   *16*
A Marshall Island sailing chart. Courtesy British Museum   *21*
Fishing with a throw net. Courtesy Hawaii Visitors Bureau   *25*
The bark *Endeavour*. From the painting by Rowland Langmaid, courtesy *Endeavour*, London   *31*
Captain Cook's monument. Courtesy Hawaii Visitors Bureau   *41*
Joseph Priestley's apparatus for experiments on air. From Joseph Priestley: *Phil. Trans. Royal Soc.*, Vol. 62, 1772   *48*
Microscopic marine diatoms. Photograph by E. J. Ferguson Wood, courtesy William M. Stephens, University of Miami   *54*
A marine diatom at 3,000× magnification. Courtesy Scripps Institution

of Oceanography, University of California, San Diego   55
The ascidian *Diazona violacea*. Courtesy Dr. G. W. Potts, Marine Biological Association of the United Kingdom   60
Forbes' lampoon of the naturalists' dredge. After Edward Forbes and Robert Goodwin-Austen: *The Natural History of the European Seas,* Van Voorst, 1859   63
The starfish *Luidia ciliaris*. Courtesy Dr. G. W. Potts, Marine Biological Association of the United Kingdom   65
Haemulon flavolineatum (Desmarest). Courtesy Professor Walter R. Courtenay, Florida Atlantic University   79
An energy conversion out of the past. Photograph by the author   91
A luminous angler fish converts chemical energy into light. Courtesy Dr. Yata Haneda, Yokosuka City Museum, Japan   92
A rare energy conversion in the sea. Courtesy B.S.I. Geological Survey and Dr. W. D. Procter   93
Pre-adult forms of a starfish. After Sir Wyville Thomson: *Phil. Trans. Royal Soc. London,* Vol. 155, 1865   97
A young *Caput medusae*. From Sir Wyville Thomson: *The Depths of the Sea,* Macmillan, 1874   100
A new starfish species, discovered by Sir Wyville Thomson. From Sir Wyville Thomson: *The Depths of the Sea,* Macmillan, 1874   104
Zoological laboratory aboard the *Challenger*. From *Challenger* Office Reports, Great Britain, 1895   110
Chemical laboratory aboard the *Challenger*. From *Challenger* Office Reports, Great Britain, 1895   110
The *Challenger* off Kerguelen. From a sketch by J. J. Wild, courtesy the University of Edinburgh   111
The Wyville Thomson Ridge. After Sir William Herdman: *Founders of Oceanography and Their Work,* Arnold, 1923   125
Woods Hole Marine Biological Laboratory. Courtesy M.B.L., Woods Hole, Massachusetts   142
Woods Hole, 1895. Courtesy M.B.L., Woods Hole   144
Scripps Institution of Oceanography. Photograph by Glasheen Graphics, La Jolla. Courtesy Scripps Institution of Oceanography   148
FLIP. Courtesy Scripps Institution of Oceanography   149
Friday Harbor. Photograph by the author   150
A giant jellyfish. From Elizabeth C. Agassiz and Alexander Agassiz: *Seaside Studies in Natural History,* Osgood, 1871   159
A rocky beach in Massachusetts. From Elizabeth C. Agassiz and Alexander Agassiz: *Seaside Studies in Natural History,* Osgood, 1871   159

Annual growth rings in coral. Courtesy David W. Knutson, Robert W. Buddemeier, and Stephen V. Smith, Departments of Chemistry and Oceanography, and Hawaii Institute of Geophysics, University of Hawaii. Cover photograph, *Science,* Vol. 177, 21 July 1972. Copyright © 1972 by the American Association for the Advancement of Science   *163*

Weight statistics for a sample of a human population. After Paul G. Hoel: *Elementary Statistics,* Wiley, 1966   *188*

Age statistics for samples of herring populations in seven successive years. After Dr. Einar Lea, quoted by Sir Alister Hardy: *The Open Sea: Its Natural History,* Houghton Mifflin, 1965   *189*

Protected outer coast. Photograph by Don T. Krogh, courtesy Pacific Gas and Electric Company   *196*

Netting plankton from the shore. Photograph by the author   *198*

Ed Ricketts collecting. Courtesy Stanford University Press   *201*

A tide pool at Pacific Grove. Photograph by the author   *208*

Tiny plants of the sea. Courtesy James Fraser, Department of Agriculture and Fisheries for Scotland, Aberdeen. From *Nature Adrift,* Foulis, 1962   *218*

Free-floating animals of the sea. Courtesy James Fraser, Department of Agriculture and Fisheries for Scotland, Aberdeen. From *Nature Adrift,* Foulis, 1962   *219*

The food chain of the herring. Significantly modified after Sir Alister Hardy: *The Open Sea,* Houghton Mifflin, 1965   *221*

An accident in the food chain. Photograph by the author   *222*

The Hardy continuous plankton recorder. Courtesy James Fraser, Department of Agriculture and Fisheries for Scotland, Aberdeen. From *Nature Adrift,* Foulis, 1962   *226*

A modern plankton recorder. Photograph by Alan Langhurst, courtesy National Marine Fishery Service, San Diego   *228*

Claude ZoBell prepares a nutrient-agar culture medium. Courtesy Scripps Institution of Oceanography   *233*

Relationships among plants, animals, and bacteria in the sea. Freely adapted from S. A. Waksman: *Scientific Monthly,* Vol. 38, 1934   *239*

Chloroplast from a bean leaf. Electron micrograph courtesy Professor T. Elliot Weier, University of California, Davis   *245*

The world's first full-scale nuclear reactor. Courtesy Oak Ridge Operations Office, U.S. Atomic Energy Commission   *247*

An autoradiograph. Courtesy Melvin Calvin, University of California   *249*

Laboratory photosynthesis in algae. Courtesy Melvin Calvin, University of California   250

The path of carbon in photosynthesis. After the figure in Melvin Calvin's Nobel Prize Lecture, Stockholm, December 11, 1962   251

The copepod *Calanus*. Courtesy Scripps Institution of Oceanography   268

The research vessel *Dolphin*. Courtesy Scripps Institution of Oceanography   269

Plankton from the California Current. Photograph by George Mattson, courtesy National Marine Fishery Service, San Diego   287

The northern anchovy. Photograph by George Mattson. Courtesy National Marine Fishery Service, San Diego   289

An ecosystems model. Courtesy Professor John J. Walsh, University of Washington. Reprinted from *Science,* Vol. 176, 2 June 1972, p. 971. Copyright © 1972 by the American Association for the Advancement of Science   300

PREFACE

This book has for its background the world's oceans, the living forms that make their homes there, and the sources of energy these organisms use to live, to get their food, and to reproduce their kind.

That is the background. More deeply the book is about man, and how he came to accumulate his store of knowledge of the seas, and his emotional ties to the sea: his awe, and wonder, and affection.

The only way that I knew to tell this story was through the lives and work of men who made significant advances in our understanding of the ocean world. There are many of them, and to do justice to the great sweep of their efforts, I have ranged from prehistoric voyages of discovery to today's most advanced science. But while there is much about science here, this is in no sense a scientific book. It is not a textbook on oceanography or marine biology, or even a history of these subjects, for they are so vast that any attempt to cover them completely would leave no room for reflection on how new discoveries have influenced man's changing attitudes. More than presenting mere information, then, I have tried to contribute to an understanding of the oceans—of the beauties of the marine environment itself as well as the experimental methods that seek to discover its secrets.

A further purpose of the book is to give the general reader enough background on marine science so that he may make judgments of his own about many of the contemporary problems of the ocean.

We hear much these days about the damage that industrial man is doing to the oceans. Some even say that they are doomed. What attitude should a responsible citizen take toward the dumping of radioactive residues and industrial wastes, heat pollution by atomic power plants, oil spills from offshore wells and broken tankers, and the rest? Should we automatically be against them all, or should we weigh potential hazards against potential benefits?

We are not likely to think intelligently about such questions unless we break them down into the many component questions that go to make them up, but having done so, we can surely subject these to analysis against a background of known facts and reasonable probabilities. It is thus my hope that some of the ideas in this book will stimulate thinking along these lines.

The reader will find no final answers here. I would not give them if I could, for I am convinced that the fate of our oceans—as indeed of our society—will, in the years to come, rest on the opinions, attitudes, and actions of informed, intelligent men.

Of course, they will act according to their own biases, but I hope that I have shown enough about the promise of modern science for the benefit of mankind (and man's disposition to use it wisely) to sustain the belief that these biases will be optimistic. For today's pressing need is surely for confident, constructive action.

No book of this range could have been written without the help of many expert hands; it is a pleasure to acknowledge those that I am consciously aware of, and also the many others who helped in more subtle ways. I owe special thanks to Joel W. Hedgpeth, resident director of Oregon State University's Marine Science Center at Newport, and Nelson Fuller, public affairs officer at Scripps Institution of Oceanography at La Jolla, for their guidance and generous gifts of time. They are, of course, in no way responsible for what I have done with the information and suggestions they provided.

Others who helped in more ways than they may realize are Donald P. Abbott of Stanford University's Hopkins Marine Station at Pacific Grove; Richard Astro of Oregon State University; Andrew A. Benson of Scripps; Robert W. Buddemeier of the University of Hawaii; Melvin Calvin, director of the Chemical Biodynamics Laboratory, University of California, Berkeley; Peter Dohrn, director of the Stazione Zoologica at Naples; Phillis M. Dowine of the University of Edinburgh; James D. Ebert, director of the Woods Hole Marine Biological Laboratory; James Fraser of the Scottish Marine Laboratory, Aberdeen; Y. Hanada, director of Yokosuka City Museum, Kanagawa, Japan; Cadet Hand, Jr., director of the University of California's Bodega Marine Laboratory; Osmund Holm-Hansen of Scripps; Bayard H. McConnaughey of the University of Oregon; G. W. Potts of the Plymouth Marine Biological Laboratory, England; Judith Reed of the Bernice P. Bishop Museum, Honolulu; Edward F. Ricketts, Jr.; William M. Stephens of the University of Miami, Florida; T. Elliot Weier of the University of California at Davis; Trevor I. Williams, editor of *Endeavour,* London; and Claude E. ZoBell of Scripps. For all those who provided photographs go thanks in addition to the usual credits that appear elsewhere.

For the contributions of my wife, Claire, only a word is needed: indispensable.

And my special thanks to Harold Strauss and his editorial staff for their astute and perceptive work.

R.R.W.

Orinda, California

INTO
THE
OCEAN
WORLD

1. MAN, SCIENCE, AND THE SEA

Man's kinship with the sea was foreshadowed long before the human race emerged. A billion years ago the living forms that preceded us were conceived in the sea, and the fluids of their cells were similar to the sea water in which they lived. After all these millions of years, the life streams that pulse in our own veins are much the same. The sea is in our blood.

The seas are also the most important physical feature of our planet. They cover nearly three-quarters of its surface to an average depth of two and a half miles; they contain by far the greatest mass of all the living things on earth; and they constitute an enormous heat engine that converts the energy of the sun into forms that we can use.

As close as we are to the sea in these biological and physical ways, we are perhaps even closer to it in our emotions. When you rest your elbows on a ship's rail and marvel at the indigo of a tropic sea, your tensions find relief, and you wait with relaxed anticipation for the silvery flutter of the next school of flying fish. A dolphin cavorts alongside, and you smile as though the two of you were sharing a mutual pleasantry. A migrating whale lumbers by, covered with barnacles, and its erratic low spout perhaps reminds you that your garden spray at home needs repairing —but you could hardly care less. The swish of foam beside the ship's prow is soft, and the trade wind plays gently about your face, and the faint cry of sea birds flying low on the horizon echoes down a million years of evolution. You and the ocean world are one.

Or you walk beside the sea at sunset, the evening sky flaming like molten steel, and as the reflecting waves ripple streamers of seaweed in and out, you cast idly about for living things along the rocky shore. As you watch quietly you see them everywhere. Here is a rock weed, and there feather boa kelp, and farther along you come to surf grass and sea palm. You begin to make out limpets, and barnacles, and mussels, and now and then a scurrying shore crab.

In a contemplative mood you begin to wonder just what the relations among all these living forms may be, among themselves and with their environment. Thus you enter into the most fascinating of all man's relations with the sea: the intellectual. You want to understand. And the passion to understand is the beginning of science.

Science has been defined in many ways. To begin with the dictionary, it is "systematic knowledge of the physical or material world." If we accept that for the moment, we will find first-rate scientists among the Polynesians of old, who had an intimate knowledge of their world long before they knew how to write it down. It is hardly surprising that this knowledge was about the sea, for beyond its horizon lay new lands—providing relief from population pressures (even then), new food sources, and, above all, new adventure.

But long, long before that, man had to emerge as a thinking, reasoning creature. Where might this have occurred? Some speculate that it happened in Africa, along the banks of some inland lake or river; others, the interior of India. More recently the noted cultural geographer Carl O. Sauer advanced the hypothesis that it took place somewhere near the sea.

"It may be, as has been thought," he wrote, "that our kind had its origins and earliest home in an interior land. However, the discovery of the sea, whenever it happened, afforded a living beyond that in any inland location.

"[My] hypothesis is that the path of our evolution turned

A man may dig clams for food; but the tang of the sea, the crash of the surf, and the peace of a lonely beach may give him richer fare.

aside from the common primate course by going to the sea. No other setting is as attractive for the beginnings of humanity. The sea, in particular the tidal shore, presented the best opportunity to eat, settle, increase, and learn. It afforded diversity and abundance of provisions, continuous and inexhaustible. It invited the development of manual skills. It gave the congenial ecological niche in which animal ethology could become human culture."

Perhaps the early scientists of the sea of whom I spoke were descendants of such a culture. In any event, they lived much before the familiar Phoenicians and Greeks who learned to navigate close to the shores of the Mediterranean Sea and the Persian Gulf. These Polynesians were of a far earlier culture. Their case has been persuasively put by the late Eric de Bisschop, who spent a lifetime studying early maritime civilizations:

"Numerous writers in many countries have written histories of navigation 'from ancient times to the present day.' They all have two points in common. First it is always the Mediterranean Sea in which prehistoric man learned his first lessons in the art of navigation . . . ; to the Mediterranean is thus reserved the honor of having produced the ancestors of all the sailors of the world.

Second, the Polynesian is either ignored or treated as though he belonged to another planet. These writers always forget that it was not until the fifteenth century that our [Western] civilization could boast its first real seaman, Christopher Columbus. We must surely admit that our culture, our knowledge, and our maritime sense were still young, in truth hardly born, when Magellan undertook his first crossing of the Pacific. Hardly born indeed when we consider that two thousand years before him the tough maritime ancestors of the Polynesians, with their rafts, double canoes, and their 'outriggers' had furrowed the oceans over more than half the world."

The Polynesians themselves spread in two great waves, widely separated in time, to explore and populate the great Polynesian triangle, which is bounded in the east by Easter Island, in the north by Hawaii, and in the southwest by New Zealand. The first migration came to an end in about A.D. 450, when Hawaii-loa sailed from Raiatea to Hawaii. The second began sometime in the eleventh century and ended about 1275 when the Raiatean high priest Paao dropped anchor in Hilo Bay, Hawaii.

During the centuries-long period between the two great Polynesian waves, a significant advance in seamanship took place in Europe. The Norsemen of Scandinavia learned two crude ways of determining latitude when they were far from the sight of land. At night they measured the angular height of the North Star, which could be done with a notched stick. At noon they measured the declination of the sun at its azimuth, using a shadow-board floating in a bowl of water. Like the Polynesians, they had no compasses, but in spite of that, they sailed to the Faeroe Islands before A.D. 800, to Iceland by 870, and to Greenland within another century.

The Norse ships were not the feared slender dragon boats of the coastbound Vikings, but rather sturdy, beamy craft called *knarrs,* skillfully fashioned of oak and pine, secured with iron rivets, and caulked with animal hairs soaked in tar. They could weather almost any North Atlantic storm, and it was in such a *knarr* that Leif Ericson, son of Eric the Red, sailed from Greenland to America in A.D. 1002.

During the two hundred years that followed the second Polynesian wave, little occurred to further man's understanding of the sea. Then suddenly, in 1492, the Age of Discovery began its thirty-year span. Columbus found the West Indies, Vasco Núñez de Balboa sighted the Pacific Ocean from a mountaintop in

Panama, and Ferdinand Magellan sailed from Spain with the hope of encircling the globe. He reached the Philippines, but was killed by natives there, and the first circumnavigation was completed by Juan Sebastián del Cano, commanding the *Victoria* of Magellan's fleet. Del Cano touched Spanish soil again in 1522, and the Age of Discovery ended.

Within still another century, modern science flowered. The English clockmaker John Harrison invented a ship's chronometer that lost only fifteen seconds during a five-month crossing of the Atlantic. Sailing from England to Jamaica in 1761, H.M.S. *Deptford* carried Harrison's chronometer No. 4, and upon arrival there the longitude reckoned by the timepiece was accurate to one-fiftieth of a degree. With such accuracy now possible, the Royal Society of London sent Captain James Cook to make extensive marine surveys throughout the Pacific.

In the same year that Cook returned to England from his first voyage, his meticulous journals packed with new observations, the English chemist Joseph Priestley laid the cornerstone for understanding photosynthesis, the process by which green plants convert the energy of sunlight directly into chemical energy. Priestley had found that green plants consume carbon dioxide and give off oxygen, and he reported his discovery to the Royal Society as a series of experiments on "different kinds of aire."

Now the biologists, especially at the beginning of the nineteenth century, gave more and more attention to the study of life in the sea. Edward Forbes undertook his surveys of life in the British seas and in the Mediterranean. Louis Agassiz founded the first marine biological laboratory in the United States. Robert Mayer showed, through physiological studies of the redness of blood in the tropics, that energy is never created and never destroyed. Sir Charles Wyville Thomson and Sir John Murray led the voyage of the *Challenger,* often called the greatest oceanographic expedition ever undertaken.

With the completion of the *Challenger* Reports as the twentieth century dawned, biologists of widely divergent interests began to turn to the seashore as a source of plant and animal materials ideally suited to the study of such varied fields as ecology, physiology, biochemistry, animal behavior, genetics, and evolution. But seaside laboratories, although critically needed, had a slow and difficult growth. In France, Italy, England, Scotland, Scandinavia, and the United States, they emerged from the most indigent beginnings to institutions that at last—nearly a century

A biology teacher and his student use a Smith-McIntyre rig to dredge marine samples from the shallow bottom of Tomales Bay, just north of San Francisco.

later—are starting to have the capabilities that their founders envisioned.

In America, at about the time that Charles Eliot became the youngest president of Harvard, Alexander Agassiz went off to the copper mines of northern Michigan, earned a fortune, and spent millions on endowments of Harvard's museum and on exploring more areas of the world's oceans, probably, than any other man. In Liverpool, Sir William Herdman brought the strands of a dozen different sciences together into one web, which was to become the modern concept of oceanography. Finally, from the Antarctic and up through the reaches of the British seas, Sir Alister Hardy studied, and pushed, and encouraged the progress and welfare of the herring fisheries.

Later, from consideration of the biologically rich littoral between tidemarks of the Pacific Ocean in California, Edward F. Ricketts wrote with keen insight of the ecology of countless marine communities, incidentally receiving some popular notice as "Doc" in John Steinbeck's book *Cannery Row*.

With the end of World War II, scientists of every kind became more and more specialized, but paradoxically it also became obvious to the entire scientific community that discoveries in a seemingly narrow field often made revolutionary contributions to entirely different fields. Thus the team approach to problems of the sea became more and more the rule. At the same time, the research tools that became available to these teams greatly surpassed—in precision and power—anything that had gone before.

The stories of these leaders of research relating to marine life are truly high adventures—adventures of brilliant minds seeking answers to the deepest mysteries of the sea. Many of the facts on which these stories are based are available in the published literature, but much of that literature is not for everyone; some of the language is abstruse, and some provides an abundance of detail of interest only to the specialist. The record of the *Challenger* expedition alone fills fifty weighty tomes. Therefore, for this book I have had to be highly selective and have chosen those topics and those men that seemed to have the greatest impact on the development of our knowledge of marine organisms, their energy relations, and their environmental preferences.

Sea creatures and their behavior are fascinating enough in their own right, but as we approach the closing quarter of the twentieth century, we find that industrial man is influencing the ecology of the ocean in ways never before seen. Conservationists have pointed out many areas—especially on land and in the air— where unthinking acts have led to tragic consequences. The oceans, on the other hand, are so vast that many biologists believe that such heedlessness has not caused irreparable damage—yet. But it surely could. Fortunately, they feel that the power to turn the tide is still within our reach.

What, then, is our most promising course of action? I believe that we have two primary lines in defense of our ocean environment. The first lies in the ongoing research that is being carried out by scientists in marine laboratories and elsewhere throughout the world. The quantity and high quality of this work are reassuring, and firmly point to the inescapable conclusion that we should encourage and support a great deal more of it. It should be richly rewarding, for as Thomas Kuhn has pointed out, normal research activity today is particularly efficient.

Where can we expect such research to lead us, and how can society best implement its results?

On a June day in 1971 I sat in the office of Dr. James D. Ebert, director of the Woods Hole Marine Biological Laboratory on Cape Cod. I had been impressed by his description of the wide-ranging public lectures that are a feature of the Woods Hole summer program, and especially by a projected series of weekend sessions on national policy in the life sciences. As we speculated on how such a national policy might be made most effective, Dr. Ebert turned to his files and drew out a copy of a presidential address that he had delivered the preceding summer before the American Society of Zoologists. As I read the title aloud, my voice must have betrayed some astonishment at the words: "The Effecting of All Things Possible." Dr. Ebert smiled. "I took that from Sir Francis Bacon," he explained. "In the *New Atlantis* Bacon wrote that *science* is the effecting of all things possible." But this aspiring view of what science can "effect" may find formidable social and political stumbling blocks in its path.

Thus the second line of defense of the ocean system lies on the social level, and surely progress there will require a fuller understanding of marine science—together with its history—on the part of every intelligent citizen.

In his presidential address, Dr. Ebert made an appeal. "I would plead," he said, "for more effective communication of what is already known. Today we are told that what we do must be relevant to society's needs. But relevant *when?* Today, tomorrow, a decade hence? Today's advances . . . will surely underlie a new technology, and will offer major new forces for the alteration of the human condition. Issues of a new dimension will surely be raised. It is crucial that these issues be made understandable to non-scientists who must participate in decision-making."

It is to that end that I have written this book. It is not in any sense a textbook, for, as Thomas Kuhn has also pointed out, "a concept of science drawn from [textbooks] is no more likely to fit the enterprise that produced them than an image of a national culture drawn from a tourist brochure or a language text." Rather, my aim has been to present the concept of marine science that emerges from the historical records of the research activities themselves—from early times to today's plans for tomorrow's work.

Such records lend themselves well to development by the narrative method, and that is what I have used, always with due regard for scientific accuracy. The adventures in discovery that follow should offer no technical difficulties to anyone. They should, moreover, provide the reader with a good background against which to make his own judgments about future social and political problems of our ocean heritage. At the very least, I hope he will find the stories enjoyable.

But in spite of his love for his homeland, Hawaii-loa was deeply disturbed. The people of Raiatea were anything but peaceful. Bitter quarrels sprang up over the ownership of land; brother killed brother over which of the Tahitian gods should be accorded highest place; and there was constant pressure to drive people out, to leave more land and food for those who stayed. Above the turmoil there sometimes arose a few who had no fear of exploring unknown seas in search of new homes for themselves and their followers, and Hawaii-loa was one of the first of these.

His qualifications were unsurpassed. He knew the nearby sea in every direction as one knows the palm of his hand. And to augment his own knowledge, his navigator, who was called Makalii after the constellation of the Pleiades, was known throughout the islands as an expert. These two had made many voyages together, especially to the eastward, to the Sea-where-the-fish-do-run. It was on one of these excursions that Makalii had urged that someday, at the time of the rising of the Pleiades, they should leave Havaiki, and set out for some far-distant land.

Here was the call to adventure. In the old way, Hawaii-loa would build a great canoe, and he and Makalii and their people would sail toward the Pleiades, using the planet Jupiter as their morning star. They would sail until they found a far new land upon which they might make a home and live in peace.

Such a voyage might last many months, and the most careful preparations would be necessary. To begin with, a double canoe must be built. Hawaii-loa would have to employ crews of skilled craftsmen for that task, and it would be his responsibility to feed and pay them. Accordingly, he commanded his people to plant extra crops, and women to make bark cloth from the paper mulberry tree, and to plait pandanus mats; and the hunters to go into the forest to catch birds, so that he would have a store of red feathers with which to make payment gifts.

Soon the expert craftsmen were at work, planning a great canoe to the smallest detail, for it must sail swiftly and be capable of riding out the roughest seas. The task that lay before them was great, and would take months of the hardest work.

First came the selection of a tree, huge and sound throughout. Deep in the mountain forest Hawaii-loa sought out a giant koa. Its crescent leaves looked healthy, but that did not tell about the soundness of the trunk. For the answer to that question, Hawaii-loa turned to one of the many signs of nature that he knew so well. Silently he and his company watched for the com-

ing of the little *elepaio* bird, for they had heard its song nearby. Soon the speckled little bird, its breast a reddish brown, hopped along a branch, and up and down the trunk. In a moment it began to peck industriously. So that tree would not do. The bird had showed them that the trunk was infested with worms.

The canoe-builders sought out other trees, and when they found one on which the *elepaio* hopped about but did not peck, they felled it with their keen stone axes. Many hands helped to drag the huge log through the forest and down to the sunny beach, where the real work of building the twin hulls would begin. Each hull would be some fifty feet long, and as strong and seaworthy as skilled hands could make it. There were, in those days, double canoes twice that length, with elaborately carved bowpieces and sternpieces, but those were for royal display in calm lagoons—not for long voyages where gale-driven winds could break all but the sturdiest craft to bits.

After the log was on the beach site, the builders took charge. Hawaii-loa had engaged master craftsmen, each of them responsible for one side of a hull. Under each master was a crew of workmen, and every workman took pride in his own skill. Each had his own set of tools, with adzes and chisels chipped with care from basaltic rock and ground to the finest edges. They varied in size and shape to fit the specific tasks for which they were designed, and they were lashed to short wooden handles with braided coconut fiber.

Whereas many of the Polynesian canoes were dug out of single logs, Hawaii-loa's was a far more ambitious project. It was to be double, with a platform between the two hulls, and a sheltering cabin mounted on the platform. And since Hawaii-loa wanted it larger than any available tree, he directed that it be assembled out of planks. This required the cutting and shaping of individual timbers for the keel, the hull, and the decks. The adzes flew fast, and soon heated to the brittle point of the basalt. The workmen cooled them by driving them into the moist trunks of banana plants, and over and over they resharpened them on blocks of sandstone.

The design of each hull was that of a carvel, with the curved

This view of the bow of an ancient Tuamotuan sailing canoe clearly shows the carvel design of the curved planks, the caulking of coconut fibers and breadfruit gum, and the braided sennit lashings.

planks laid edge to edge. They did not fit perfectly at first, but the workmen smeared the upper edge of each lower plank with thin mud, and spots that needed trimming down were shown by the areas of transfer. As a modern dentist adjusts a gold crown to the irregularities of the opposing tooth using carbon paper, so the craftsmen trimmed their rising tiers of planks to perfect fits. Even so, the seams had to be caulked, and for this the workmen beat together a mixture of coconut fiber and breadfruit gum.

For fastenings they had no nails or screws, for no metal was known to them. To lash the planks together they used braided coconut sennit. One such strand is not strong, but many lashings would produce a seaworthy craft. The workmen bored closely spaced holes near the edges of adjoining planks, through which to lash the sennit. Laboriously they pierced these heavy planks with the pointed shells of the *terebra,* known as the auger shell, and often they worked these to the nub.

When all the lashings were firm, and the caulking set, the hull was scrubbed out with fresh water, left in the sun to dry, and then treated both outside and inside with a slurry of red earth and charcoal. Finally, the booms were lashed between the two hulls to hold them firmly together, a deck of straight planks was laid over the booms at right angles to them, and the deckhouse was erected there, centered between the two hulls.

At last came the launching. This was celebrated with elaborate ceremony, and the entire population of the district came, garlanded with flowers and sweet herbs, and clothed in their best finery. They feasted, and drank *kava,* and offered up prayers to the gods of the sea, for to them the sea was the altar of the gods.

Finally, the ship was launched. Workmen quickly placed smooth round logs between the canoe and the sea, all hands pushed, and the vessel rolled down the skids until it floated gracefully in the lagoon.

With the ship launched, Hawaii-loa gave orders for its fitting out. A tall mast was mounted on the deck, and that was rigged with a triangular sail of plaited pandanus mats, carefully sewn together. Paddles, bailers, and stone anchors completed the fitting of the ship itself, but that was far from the end of the necessary preparations.

For food for the long voyage, ripe pandanus drupes were picked, grated into coarse flour, cooked, dried, and finally rolled into bundles in dried pandanus leaves. Other edibles, both cooked and raw, were also dried: breadfruit, sweet potatoes, coconut meat, and shellfish.

Supplies of fresh water were carried in coconuts, in gourds, and in stoppered lengths of bamboo. Since the voyage would be long, Hawaii-loa could also look forward to collecting rain water from time to time.

After these long and careful preparations, a night finally came when the constellation of the Pleiades rose in the darkened sky, and Makalii the navigator said that the time to start the voyage had arrived. They would leave at dawn.

As the first tint of rose flushed the eastern sky, Hawaii-loa had two captive frigate birds put on board, and a pig, and a pair of chickens for breeding stock. Then all hands boarded, and the paddlers dug their blades into the lagoon. Facing forward, they could all see what lay ahead, and Makalii trimmed his steering paddle so that the bows of the canoe pointed straight to the white water where the open ocean poured through a treacherous break in the reef.

The beat of the paddles rose furiously, for it would take all their strength to make headway through the break. To crash against the jagged coral on either side would mean disaster. Swiftly through the white water they cut, as the waves of the open ocean mounted higher and higher. At last, with a shuddering pitch of the bows, the tallest of the waves was mounted, and the craft rode free. Hawaii-loa and his voyagers were on the open ocean.

But the men who dipped the bailers were now urgently driven to their task. Not only had the twin hulls shipped much water in their passage through the breakers, but also, in spite of the countless lashings of the sennit braid and the careful caulking, there had been some working of the planks in the rolling torrent, and enough leaks had sprung so that bailing would be a continuous task from that time on.

Although he could not know it—for man had never taken this route before—Hawaii-loa had set out upon a voyage of twenty-four hundred miles by the great circle route, and much farther in actual distance sailed. He did not reckon his progress in miles, however. The Polynesians always thought of distance in terms of time: the distance from one island to another would be given in canoe-days. The voyage on which Hawaii-loa embarked was to last for more than five canoe-months.

Nearest to the methods of the modern navigator with his

sextant and chronometer was Hawaii-loa's knowledge of the stars, together with his innate time sense, which he had developed into an instrument of remarkable precision. He knew a hundred and fifty stars by name, and the fact that they stayed in the same relation to one another as they moved across the sky from dusk to dawn. He knew which star would stand directly above a particular island, and that a certain star that lay to the north of that one would pass over another island also farther to the north. By visualizing the relations among these guide points in his mind, he possessed an accurate mental map of all the islands that he knew.

How could an early seaman, possessing no instruments but his own keen senses, have been able to tell when a particular star was directly overhead? The late Harold Gatty has given us a clue. He found that by looking directly up while slowly turning his body in a circle, he could pinpoint a star at the zenith within one degree of arc. That would fix his position on earth to an accuracy of sixty miles or less—well within the range of other direction-finding methods that were known to the Polynesians.

Hawaii-loa had also observed that, while each star kept to its own path, it rose in that path somewhat earlier each night. We do not know just how accurate the Polynesian time sense was, but since stars rise four minutes earlier from one morning to the next, it would take a truly amazing inner clock to detect that difference in a span of twenty-four hours.[1]

As we have seen, Hawaii-loa's navigator steered toward the Pleiades, or the Seven Little Eyes. That was indeed long ago, for the seventh sister of the Pleiades has not been visible to the unaided eye for centuries. Another *mele* relates that Makalii held his course in the direction of Orion's belt. No doubt both were consulted, along with many other stars and constellations well known to the Polynesians.

Hawaii-loa also followed the trade winds. Havaiki lies some seventeen degrees south of the equator, and in those southern latitudes the trade winds blow steadily from the southeast. This was a daytime guide, when the stars were not visible, and it also helped to make speed, for the voyagers were heading north, and they could sail quartering the wind.

But when they came within some five degrees of the equator, a new and unfamiliar thing confronted them. The reliable south-

---

[1] I have written elsewhere about the remarkable biological clocks of plants, animals, and men; see my *The Living Clocks* (New York: Alfred A. Knopf; 1971).

Instead of carrying navigational aids in their heads as Hawaii-loa did, the Marshall Islanders made physical chart models from the ribs of coconut palms lashed together with coconut fiber. The curvature of these ribs accurately represented the curvature of the dominant swells around the islands. The islands themselves were represented by the small shells mounted on the ribs.

east trade wind fell off and finally died. A dead calm lay upon the sea. They had entered the region of the doldrums, and although they could not know it, they would have to travel six or seven hundred miles northward before they would encounter a wind strong enough to drive the canoe. For days on end the sail hung lifeless against the mast.

Now all hands turned to the paddles, and to the steady beat of the ancient canoe chants, flashing blades dug the water and on the return stroke clashed against the gunwales.

This tedious passage through the doldrums took nearly half the time of the entire voyage. The paddlers sweltered under the equatorial sun; they sweated rivers, and would have drunk three times their water rations if Hawaii-loa had permitted it. But at last the ordeal ended. A day dawned in less unbearable heat, a breeze sprang up, and Hawaii-loa's crew shipped their paddles in wonder at this new trade wind. For it blew in the wrong direction. Fresh and strong, now, as the trade wind they knew so well, it blew from the northeast. To steer toward Orion's belt or the Pleiades, they would have to tack into this wind. But they were skilled in tacking. Often enough in earlier times they had sailed to islands south of Havaiki against the southeast trade wind. And they had made a new discovery about the ocean and its winds. Hard-won lore, it would find its place in new chants that would be passed on, to guide other voyagers in generations yet to come.

In addition to the stars and the winds, Hawaii-loa had living aids for his long-distance navigation. He knew that the golden plover flew in a north-south line in its annual migration. He also knew that the same line was followed by the rarer bristle-thighed curlew. Probably neither he nor Makalii would have considered taking the voyage if they had not been familiar with the migrations of these birds, and drawn the conclusion that other lands must lie along their paths of flight.

Now, after more than four months on the open ocean, Hawaii-loa took careful inventory of his remaining stores. True, the crew had been on short rations for some time; in fact, they had trained themselves to a short-ration routine before they had left home. But now Hawaii-loa could clearly see that even on the shortest rations his men could not last another month.

It was time to call on his shorter-range navigational aids. First he sent aloft one of his captive frigate birds. It rose on its enormous black expanse of wing, its white breast shining in the sun, its long forked tail opening and closing like scissors. This most skillful of all sea birds rose higher and higher, and after wheeling to scan the ocean, settled into swift flight to the northward. If it did not return that day, Hawaii-loa would know that land lay not more than a few canoe-days away, for the frigate bird will not spend a night upon the water. At rest he is a land bird, for his feathers become dangerously waterlogged if he floats for very long in the water. Therefore, if Hawaii-loa's frigate bird did not sight land, he would almost certainly return to the canoe. This is just what happened on this first release, and Hawaii-loa knew that the nearest land lay several canoe-days beyond.

Even so, he began a more careful study of the patterns of the swells that surged on the surface of the ocean. As long as they moved in a straight-line wave form, he would know that no island was near at hand. But if his close watch should detect the beginnings of a curving pattern, he would know that land lay not far distant. As ocean waves approach an island, or even a low-lying atoll, they form a great curving arc with the island near its center. At the same time, on the lee side of the island, counterswells curve in the opposite direction. From his long familiarity with these patterns, Hawaii-loa could detect a distant island from even farther away than he could see the reflection of a distant lagoon on the lower surface of a cloud.

Every man in the canoe also kept a sharp watch across the sky, for each hoped to be the first to see a tropic bird. It was Makalii the navigator who had that honor. Keeping one hand on the steering paddle, he raised the other with an exultant cry. The white wings of a tropic bird flashed through the early morning mist. It was joined by another, and another. As the outlines of the approaching birds became clear, Hawaii-loa smiled his relief, for these graceful birds were a sure sign of land. As they drifted closer he could make out the slender tails, longer even than the bodies, and the dark patches above the eyes and on the upper surfaces of the wings.

At once Hawaii-loa released another frigate bird. It rose swiftly, wheeled, and without a moment's hesitation set an unswerving course to the north. It never returned, for it had sighted an island dead ahead.

Makalii the navigator pointed to the pig. It was snuffling loudly and had grown very restless. This was still another sign that the voyagers had looked forward to. Pigs have a most acute sense of smell, and can detect land odors over great distances at sea. At the first faint whiff, they become so restless as to leave no doubt that land lies near.

The Polynesians themselves had a sense of smell that was

keen enough, and before the sun had set that day, Hawaii-loa caught a faint fragrance that he was to love for all his days. It was the sweet fragrance of the *maile* vine that grows down the slopes of the mountains that he was nearing.

The warm night passed slowly, for after five months of lonely sailing, visions of land so near kept the company restlessly astir. And as dawn finally broke, they saw a long white cloud in the distance; it did not drift with the trade wind, but was quite stationary. It was woolier than the usual trade-wind clouds, and past it on a lower level drifted smaller clouds. The pattern was a sure sign of land close by.

Now, as the sun mounted in the sky, three high mountain peaks lifted above the standing cloud. The tallest was capped with white, and Hawaii-loa named it Mauna Kea, the white mountain. The next was not quite so high, but its ridge stretched far, and he called that Mauna Loa, the long mountain. The third, north and west of the other two, he named for his wife, Hualalai, for whom he was to return on a second voyage. The new island he named after himself—Hawaii. These are the names they bear to the present day.

Sailing toward the lee shore of that island, Hawaii-loa saw that they were reaching shallower water, for from the deep blue of the open ocean, the color changed to turquoise, and then to green, and very close to the shore it was a sandy yellow. When he saw the great coral heads projecting above the white-breaking surf, Hawaii-loa ordered that the sail be lowered and that the paddlers take over. To stay clear of the coral heads, he held the canoe's course to water of the darkest color, through a break in the reef, until finally the twin hulls beached on the yellow sand with a reassuring crunch. The mariners tumbled out and luxuriated their bare brown feet in the sand. After five months afloat, a new land stood firm beneath them. They were the first men to touch this shore, and they stayed to establish their permanent home there.

Not long after this first settlement of the islands, communication between Hawaii and the rest of Polynesia came to an end, and it was not until many generations had passed that a new group of navigators made their way northward.

Late in the thirteenth century the high priest Paao sailed

Hawaii-loa and his followers were highly skilled at fishing with the weighted throw net. Today's Hawaiians have retained these skills.

from Raiatea in a double canoe similar to Hawaii-loa's, and like Hawaii-loa, he came at last to the island of Hawaii. But he approached from the windward side, and came to anchor in Hilo Bay. Sir Peter Buck places the date at 1275, which confirms evidence from other sources that the Polynesians sailed regularly from the Society Islands to Hawaii for a period of at least two hundred years. It was also during this period that they sailed twenty-six hundred miles southwest from the Societies to New Zealand, discovering and settling that Land-of-the-long-white-cloud.

What happened on Easter Island is not so clear, but it would not add much to the picture if we knew it. The essential point is that we have a detailed view of one of man's earliest and greatest maritime civilizations, which could have come about only through intimate familiarity with nature, and the use of that knowledge to guide unprecedented voyages of exploration.

Two hundred years after Paao dropped his stone anchor in Hilo Bay, the navigators of the West began the Age of Discovery, which opened with Columbus and closed with Magellan in 1522.

Throughout those centuries the Hawaiian culture grew and flourished, and all of the eight islands of the chain were populated. But still another two hundred and fifty years were to pass before a Western explorer "discovered" those islands. According to the Hawaiians, that happened in this way.

Lono, the Polynesian god of peace and the harvest, had first been manifest in Tahiti, but on one of the voyages that followed Hawaii-loa's, he came to Hawaii and lived there for many years, to the great benefit of the people. Although he later went back to Tahiti, the Hawaiians believed that one day he would return and take his rightful place among the Hawaiian tutelary gods. That he did, on the morning of Sunday, November 30, 1778.

That morning incredulous messengers burst upon the Hawaiian king and told him in awed tones that two huge *heiaus,* or godly temples, had been seen floating in the western sea. They could be no other than the temples of the god Lono, returned at last from Tahiti, for from the tallest masts ever seen there fluttered huge *makahiki* banners—Lono's banners of the festival of the new year.

At that very moment the captain of the smaller ship raised his spyglass and, with a seasoned mariner's eye, surveyed the black coral crescent of Kealakekua Bay. Lono had returned, in the person of an officer of the British Navy, Captain James Cook.

3. COOK: AROUND THE WIDE PACIFIC

On Friday, August 25, 1768, Captain James Cook, commanding the bark H.M. *Endeavour,* weighed anchor off Plymouth, England, and set sail for the South Pacific. Two hundred years later, in London, the Hakluyt Society marked the date with a series of celebrations in tribute to Cook's enrichment of the world's knowledge.

The world regards Cook as among the great adventurers and explorers—and rightly so. What is not so commonly recognized is that Cook also belongs among the world's great scientists—with Sir Isaac Newton and Charles Darwin, although his contributions were of a very different sort. Newton and Darwin were able to derive far-reaching general laws from their own observations. Cook was not. He was a participant, rather, in the great collecting phase of the eighteenth century, during which facts about the natural world were accumulated so rapidly that they could not be digested at the time. It remained for later specialists with sharp new techniques to bring out the full implications of the information that Cook amassed.

What manner of man was this? The journals of those who sailed with Cook, and knew him intimately, together with an even casual inspection of his own voluminous journals, show a markedly scientific turn of mind. He was cautious, careful, thorough, studious, quiet, reticent, even self-deprecating—and beneath all of these qualities flamed a quick, incisive imagination.

We know his physical appearance from a portrait that now hangs in the National Maritime Museum in Greenwich, painted in 1775 by Nathaniel Dance, R.A. According to David Samwell, a surgeon to Cook, this was the only portrait that bore any likeness to him. Mrs. Cook liked the Dance portrait too, although she thought it made him look too stern.

Dance painted Cook sitting, holding a navigational chart on his right knee, with his face turned somewhat to the left, as though he were about to explain some point in navigation. His eyes are direct and serious, and his mouth is firm but certainly not harsh. The brown hair, in the style of the period, begins above a high, intelligent forehead. The coat is dark blue, the sleeves heavily trimmed with gold, out of which flow white ruffled cuffs. Neither hand shows any sign of the maiming that occurred in an earlier powder-horn explosion. The coat is faced in gold and white, and the white waistcoat carries a row of closely spaced gold buttons. It is a convincing portrait of a man who, at the age of forty-six, had developed in so many different scientific fields that he had been elected to membership in the Royal Society and had been presented to the King.

It seems an unlikely portrait of a man who was the second of seven children of a day laborer in eighteenth-century England. Cook was born on October 27, 1728, in a tiny two-room clay cottage in the village of Marton-cum-Cleveland.

Young Cook's education was spotty, and he must have made the most of what it offered. We know that he showed a "remarkable facility in the science of numbers." Still, he worked as a farm boy until he was seventeen, and then was sent to learn the "mysteries" of the grocer's trade in the fishing port of Staithes.

That lasted only eighteen months. Dr. A. Grenfell Price has conjectured that since Cook then lived within sight and sound of the sea, with ships constantly passing before him, he succumbed to the age-old lure. At any rate, in July 1746, when he was not yet nineteen, he was apprenticed to John and Henry Walker, shipowners of Whitby, who plied the British coal trade. For the next ten years Cook learned at first hand about the sailing problems of the North Sea, and when he was ashore he studied mathematics and navigation with steady diligence.

When that decade ended, the Walkers offered Cook command of one of their ships, but although it was a well-merited recognition of his maturing competence, Cook decided to decline. England was then on the brink of the Seven Years' War, and Cook thought his duty lay in one of His Majesty's ships of the line. Soon after his enlistment he was promoted to master's mate of the *Eagle* under Captain (later Sir Hugh) Palliser. In this capacity he was responsible for a wide range of duties, especially in navigation, and by 1764 he was given command of the schooner *Granville*. During the next three years he made charts of Newfoundland and Labrador, which were so accurate that they were not completely superseded for more than a hundred years. These, and his observations of the eclipse of the sun in August 1766, were brought to the attention of the Royal Society. That body was well impressed, as it was also with the commendation of Cook's commanding officer, Lord Colville, who wrote that he "was well qualified for the work performed, and for greater undertakings of the same kind."

Meanwhile, in 1763, the Treaty of Paris brought peace among Great Britain, France, and Spain, and Great Britain emerged as the greatest colonial, commercial, and naval power in the world. If open war was at an end, however, economic competition among the three sea powers remained intense, and British science and government joined hands to strengthen the nation's hold upon the seas. They faced technical problems of extraordinary difficulty. Among these were the following:

- No one knew how to determine the longitudinal position of a ship at sea accurately. Many islands had been discovered, only to be lost again, because their longitudes were only vaguely known.
- No one knew how to prevent the appalling death rate from scurvy that plagued the long sea voyages of the time. When Commodore Anson took his three ships around the world between 1740 and 1744, he lost 626 out of 961 men to scurvy alone.
- It was generally believed that a great undiscovered land mass must lie somewhere south of Tierra del Fuego—the storied "Terra Australis Incognita." The known expanse of land north

of the equator was very great, and geographers reasoned that a vast southern continent must counterbalance it. Otherwise, "what would keep the earth from tipping?"

- The size and shape of Australia and New Zealand were very poorly known.
- Even less was known about the geography of the North Pacific and of the Arctic regions that adjoin it.

These were problems that the England of King George III was determined to solve, and to solve them it looked for a method and a man. The method was the reasoned scientific approach that was coming into its own throughout the Western world. The man was James Cook.

The first of these problems, that of accurately fixing the position of a given point on the surface of the earth, is the most basic. Let us look at it in some detail.

To fix a point, say that of a tiny island in the open ocean, it is necessary to know only two things: its latitude and its longitude. That is, if we know these two lines exactly, then the point at which they intersect will be known exactly also, and that will state the position of that island on the surface of the earth. This is simple enough in principle, but in practice it was extremely difficult because of the longitude problem.

Today it seems astonishing, but as late as 1740 the German astronomer and mathematician Johann Doppelmayr estimated that not more than 116 places on the face of the earth had been accurately located. Imagine, then, the problems involved in mapping, say, the coasts of Australia and New Zealand, separated even at their closest points by some thousand miles of the Tasman Sea.

Lloyd A. Brown has recently put the difficulties of world cartography very well. "There is only one way to compile an accurate map of the earth and parts thereof, and that is to go into the field and survey it; likewise, there is only one way to establish a series of base lines from which such surveys can be made, and that is to send out expeditions to the far corners of the earth equipped with astronomical apparatus and surveying instruments. This represents a tremendous undertaking requiring time, centuries of it, and money. It also requires an incentive. The nations that have been most interested in the establishment of colonies and a world trade have contributed more than others to the establishment of a science of cartography."

Captain Cook's bark Endeavour, *as painted by Rowland Langmaid. This representation is based on all the available authentic data, including the original profile plan of the vessel.*

Great Britain was a leader among such nations, and it was a natural consequence that the British Admiralty and the Royal Society chose James Cook, mathematician and navigator, to command the bark H.M. *Endeavour* on a three-year voyage of world exploration. From Plymouth, on August 25, 1768, Cook set sail into the western sea. He was then not quite forty years of age.

According to the public announcement, Cook's principal objective was to be the South Pacific island of Tahiti, where he was to observe the transit of Venus, predicted by the Astronomer Royal to occur on June 3, 1769. The Royal Society hoped that a new observance of the time it would take for Venus to cross the face of the sun would give a more accurate figure for the distance between the sun and the earth. It was reasonable for the public to accept Cook's mission as mainly scientific, for among the ninety-four persons aboard the *Endeavour,* eleven were scientists, including the well-known Joseph Banks.

It was also generally known that the transit of Venus was to be observed by astronomers from several points on earth. That Cook's observations of the event turned out to be imperfect is of little import now. What is significant is the way in which he went about his work. That is best shown in his own words. His journal for Saturday, June 3, 1769, reads:

> This day prov'd as favorable to our purpose as we could wish, not a Clowd was to be seen the whole day and the Air was perfectly clear, so that we had every advantage we could desire in Observing the whole of the passage of the Planet Venus over the Suns disk: we very distinctly saw an Atmosphere or dusky shade round the body of the Planet which very much disturbed the times of the Contacts particularly the two internal ones. Dr Solander observed as well as Mr Green and my self, and we differ'd from one another in observing the times of the Contacts much more than could be expected. Mr Greens Telescope and mine were of the same Mag[n]ifying power but that of the D^r was greater than ours. It was ne[a]rly calm the whole day and the Thermometer expos'd to the Sun about the middle of the Day rose to a degree of heat (119) that we have not before met with.

During this stay in Tahiti, Cook also showed a special flair for dealing with certain characteristics of the natives which had angered earlier explorers and which they had tried to handle with the crudest kind of cruelty.

To these Tahitians theft was simply a game, to be played with as much skill as one could command, and that was a great deal. One night while Cook lay in his bunk—fully awake, he thought—someone stole his stockings from directly beneath his pillow. That culprit was never caught, but there would surely be others, stealing other things, and Cook met the challenge with a psychological attack.

Whereas earlier explorers treated captured culprits by flogging, shooting, or hanging, Cook let it be known that anyone caught stealing would have his head shaved. A Tahitian, it seemed, would rather run the risk of hanging than be made the object of such ridicule.

Now, with the transit of Venus fully logged, Cook took out the secret sealed orders that had been placed in his hands in Plymouth and opened them. Long, detailed, and stiffly formal, they ordered him to seek out and discover the vast southern continent and, when he found it, to "take possession of Convenient Situations in the Country in the Name of the King of Great Britain."

The "Terra Australis Incognita" had always been imagined as rich and fertile, and if that proved so, George III wanted it firmly in his grasp.

On July 31, 1769, Cook weighed anchor and began his search. West of Tahiti he found and charted a new group of islands that he named the Society Islands (after the Royal Society). Continuing south as far as 40° without sign of anything but open ocean, he then turned west again and finally made a landfall. Could it be the "Terra Incognita"? As he charted in his usual fashion, he slowly came to realize that this was not a new continent, but rather New Zealand, which had already been found in 1642 by Tasman. Cook sailed completely around North Island, charting as he went. Then all around South Island also. And of course in doing so he found that a strait lay between the two islands, rather than a bay, as Tasman had thought.

That done, Cook might have been well justified in turning home. But the curiosity that drives a man of this temperament was not to be put down. Before him lay a golden opportunity to sail up the eastern coast of Australia, which was then completely unknown, and chart that shoreline. Cook seized the opportunity and headed straight for the edge of disaster:

MONDAY 11TH JUNE [he wrote in 1770]. Wind at ESE with which we steer'd along shore NBW at the distance of 3 or 4 Leagues having from 14 to 10 & 12 saw two small Islands in the offing which lay in the latitude of 60°0′ S and about 6 or 7 Leagues from the Main. At 6 oClock the northermost land in sight bore NBW½W and two low woody Islands which some took to be rocks above water bore N½W. At this time we shortened sail and hauled off short ENE and NEBE close upon a wind. My intention was to stretch off all night as well to avoid the dangers we saw ahead as to see if any Islands lay in the offing, especially as we now begin to draw the Latitude of those discover'd by Quiros which some Geographers, for what reason I know not have thought proper to tack to this land, having the advantage of a fine breeze of wind and a clear moonlight night. In standing off from 6 untill near 9 oClock we deepened our water from 14 to 21 fathom when all at once we fell into 12, 10 and 8 fathom. At this time I had every body at their stations to put about and come to an anchor but in this I was not so fortunate for meeting again with deep water I thought there could be no danger in standing on. Before 10 oClock we has 20 and 21 fathom and continued in that depth untill a few minutes before 11 when we had 17 and before the Man at the lead could have another cast the Ship struck and stuck fast. Emmidiately upon this we took in all our sails hoisted out the boats and sounded round the Ship, and found that we had got upon the SE edge of a reef of Coral rocks having in some places round the Ship 3 and 4 fathom water and in other places not quite as many feet, and about a Ships length from us on our starboard side (the ship laying with her head to the NE) were 8, 10 and 12 fathom. As soon as the long boat was out we struck yards and Topm[ts] and carried out the stream Anchor upon the starboard bow, got the Costing anchor and cable into the boat and were going to carry it out the same way; but upon my sounding the second time round the Ship I found the most water a stern, and therefore had this anchor carried out upon the Starboard quarter and hove upon it a very great strean which was to no purpose the Ship being quite fast, upon which we went to work to lighten her as fast as possible which seem'd to be the only means we had left to get her off as we went a Shore about the top of high-water. We not only started water but throw'd over board our guns Iron and stone ballast Casks, Hoops staves oyle Jars, decay'd stores &c[a], many of last articles lay in the way at coming at heavyer. All this time the Ship made little or no water. At a 11 oClock in the AM being high-water as we thought we try'd to heave her off without success, she not being a float by a foot or more notwithstanding by this time we had thrown over board 40 or 50 Tun weight; as this was not found sufficient we continued to Lighten her by every method we could think of. As the Tide fell the Ship began to make water as much as two Pumps could free. At noon she lay with 3 or 4 Strakes heel to Starboard. Latitude Observed 15°45′ South.

TUESDAY 12TH JUNE. Fortunately we had little wind fine weather and a smooth Sea all these 24 hours which in the PM gave us the oppertunity to carry out the two bower Anchors, the one on the Starboard quarter and the other right a stern. Got blocks and tackles upon the Cables brought the falls in abaft and hove taught. By this time it was 5 oClock in the pm, the tide we observed now began to rise and the leak increased upon us which obliged us to set the 3rd Pump to work as we should have done the 4th also, but could not make it work. At 9 oClock the Ship righted and the leak gain'd upon the Pumps considerably. This was an alarming and I may say terrible Circumstance and threatened immediate destruction to us as soon as the ship was afloat. However I resolved to resk all and heave her off in case it was practical and accordingly turn'd as many hands to the Capstan and windlass as could be spared from the Pumps and about 20' past 10 oClock the Ship floated and we hove her off into deep water having at this time 3 feet 9 inches water in the hold. This done I sent the Long boat to take up the stream anchor—got the anchor but lost the Cable among the rocks, after making this turn'd all hands to the Pumps the leak increasing upon us. A mistake soon after happened which for the first time caused fear to operate upon every man in the Ship. The man which attend[ed] the well took ye depth of water above the ceiling, he being relieved by another who did not know in what manner the former had sounded, took the depth of water from the outside plank, and difference being 16 or 18 inches and made it appear that the leak had gain'd this upon the pumps in a short time, this mistake was no sooner cleared up than [it] acted upon every man like a charm; they redoubled their Vigour in so much that before 8 oClock in the Morning they had gain'd considerably upon the leak. We have now hove up the best bower but found it impossible to save the small bower so cut it away at a whole Cable. Got up the fore topmast and fore year, warped the Ship to the SE and at 11 got under Sail and Stood in for the land with a light breeze at ESE, some hands employ'd sowing ockam wool &c^a into a lower Studding sail to fother the Ship, others employed at the Pumps which still gain'd upon the leak. . . . The ledge of rocks or Shoal we have been upon lies in the Lat^{de} of 15° 45' and about 6 or 7 leagues from the Main land. . . . At Noon we were about 3 Leagues from the land and in the 15° 37' South, the northermost part of the Main in sight bore N 3° west and the above Islands extending from S 30° E to South 40° E, in this situation had 12 fathoms water and sever^l Sand Banks without us. The leak now decreaseth but for fear it should break out again we got the sail ready fill'd for fothering. The manner this is done is thus, we Mix ockam & wool together (but ockam alone would do) and chop it up small and then stick it loosely by handfulls over the sail and throw over it sheeps dung or other filth. Horse dung for this purpose is the best. The sail thus prepared is hauld under the Ships bottom by ropes and if the place of the leak is uncertain it must be hauld from one part of her bottom to a nother untill the place is found where it takes effect; while the sail is under the Ship the Ockam &c^a is washed off and part of it carried along with the water into the leak and in part stops up the hole. Mr Munkhouse one of my Midshipmen was once in a Merchant ship which sprung a leak and made 48 inches of water per hour but by this means was brought home from Virginia to London with only her proper crew, to him I have the deriction of this who exicuted it very much to my satisfaction.

When Cook beached the *Endeavour* at the northeast corner of Australia and examined her leaks, he found: "Fortunately for us the timbers in this place were very close, other ways it would have been impossible to have saved the ship and even as it was it appear'd very extraordinary that she made no more water than she did. A large piece of Coral rock was sticking in one hole and several pieces of the fothering, small stones, sand &c^a had made its way in and lodged between the timbers which had stoped the water from forcing its way in in great quantities."

Of this crisis the Australian authority A. Grenfell Price feelingly said, "it may be noted . . . that a large piece of coral was blocking one of the holes in the *Endeavour*. The casual as well as the causal creates history and but for this piece of coral there might have been no British colonization of Australia."

On the other hand, when we look at Cook the man, the knowledge and resourcefulness that he displayed seem far more meaningful than this particular piece of good fortune.

After the *Endeavour* had been given temporary repairs, Cook sailed her around Cape York, through Torres Strait, and on to Batavia (now Djakarta), where he put in for provisions and permanent repairs. Within ten weeks he sailed for home, arriving in London on July 13, 1771. One of the most famous voyages in history was at an end.

The accomplishments of Cook the explorer are obvious. He had shown beyond doubt that, if a great southern continent existed, it lay much farther to the south than had been supposed. He "proved," says A. Grenfell Price, "and depicted on splendid charts that New Zealand consisted of two large, fertile and alluring islands which obviously offered enticing prospects for colonisation. Most important of all, however, he showed that, although the Dutch land of New Holland (Australia) was separated from Quiro's Espiritu Santo and from New Guinea by Torres' long-forgotten Strait, it was a country of continental dimensions; with a vast eastern coastline which appeared continuous from Point Hicks to Cape York."

Remarkable about the accuracy of Cook's charting on this first voyage was the fact that he still had no ready way of determining longitude. He had no accurate chronometer, and the calculations that he did make from Nevil Maskelyne's tables of lunar distances took laborious hours. But they were accurate. His calculation of the longitude of Port Venus was in error by only one degree, corresponding to about a mile.

The accomplishments of Cook the observant scientist were perhaps less spectacular, but surely no less significant. Everywhere he landed he studied the land, the plants, the animals, and the people; and he noted their beliefs and their tools and artifacts with a keen and discerning eye. All these he recorded in careful detail in his journals. His scientists and artists took copious notes and made many drawings, and at Tolaga Bay, New Zealand, Cook tried his own hand at pen drawing. The Royal Society was enriched with records of coral trees in Tahiti, double canoes in Raiatea, breadfruit trees and palms and temples in Huahine, tapa cloth designs in Tahiti, and fishhooks, axes, needles, pestles, and musical instruments from many exactly recorded islands. The artist Parkinson made beautiful drawings of a butterfly fish from a coral reef, and of a lobster krill, of breadfruit and the New Zealand honeysuckle, and a rough sketch of a kangaroo that he saw on Endeavour River. A treasure indeed for leisurely study by London scientists.

Cook's stay in England lasted just a year. The Admiralty saw signs of increasing activity by both the French and the Spanish in the Pacific. Quickly two new ships, the *Resolution* and the *Adventure,* were outfitted much more completely than the *Endeavour* had been, and with Cook in command of the *Resolution* and Captain Tobias Furneaux in command of the *Adventure,* they sailed, again from Plymouth, on July 13, 1772. This time they were to sail the eastern route. They made first for the Cape of Good Hope, and then south as far as the Antarctic ice allowed, for the colonial objective again was the great southern continent. Again, of course, they did not find it, but they sailed what mariners still call the "filthy fifties" in search of it, and penetrated the Antarctic Circle twice.

Cook's scientific objectives were two: to find out whether John Harrison's new chronometer would actually serve to determine longitude accurately at sea; and to show more conclusively that proper diet would prevent scurvy on long voyages. In both, he was outstandingly successful.

The chronometer that Cook carried was not made by Harrison himself, but was a copy, which the Board of Longitude had commissioned Larcum Kendall to make, of Harrison's famous No. 4. It looked like a very large watch, five inches across, with a pendant at the top as though it were to be worn like a pocket piece. Like Harrison, Kendall did not mount his chronometer on gimbals, but laid it in a plain box lined with a soft cushion. Throughout the entire voyage of three years and eighteen days, it kept astonishingly good time. Its dependability in determining longitude was fully acknowledged by Cook, who wrote on his return:

> SATURDAY 29TH JULY [1775] we made the Land about Plymouth, Maker Church at 5 o'Clock in the afternoon, bore N 10° West distant 7 leagues, this bearing and distance shew that the error of Mr Kendall's Watch in Longitude was only 7′ 45″, which was too far to the West.

Cook's preparations to combat scurvy were characteristically thorough. As antiscorbutics he took "Malt, Sour Krout, Salted Cabbage, Portable Broth, Saloup [an herb drink], Mustard, Mermalde of Carrots, and Inspisated Juce of Beer." The last-named was simply beer that had been boiled down to a tenth of its original volume to save storage space; for use, it was diluted back with water. Rations of all these were regularly issued throughout the voyage, together with fresh fruits and vegetables at each port of call. Cook also insisted that his men keep their bodies and their clothing clean—a rare thing for seamen in those days. The result was that not a single man died of scurvy—a remarkable record compared to the 33 per cent death rate usual in other long voyages of the time.

At home once again, Cook was showered with honors. He was presented to King George III, he was made a Fellow of the Royal Society, England's highest scientific honor, and he was given the sinecure of post-captain at His Majesty's Royal Hospital for

Seamen at Greenwich at £200 per year with perquisites. But Cook chafed at the inactivity, and the sinecure lasted only a year. New discovery, new adventure, and final tragedy lay ahead.

On July 12, 1776, Cook sailed for the last time from Plymouth. His chief mission was to discover the Northwest Passage, so that British shipping might proceed into the Pacific by a much shorter route than around South America. On his way to the North Pacific he cruised the South Pacific, with which he was so familiar, for eighteen months, revisited Tonga and Tahiti, and, twenty-four hundred miles to the north of Bora Bora, he discovered Hawaii. This group he named the Sandwich Islands, after the Earl of that name.

Cook's first Hawaiian landing was on the northern island of Kauai, at Waimea. There he went ashore on three different occasions, and traded nails and old iron for pigs, chickens, yams, and taro; he also took on a fresh supply of water. He visited the smaller island of Niihau, and sent a party ashore to trade a ram and two ewes, along with a variety of seeds, for salt and more yams. The trading party was forced to stay on the island for two days and nights because of the high surf, and it is Professor Ralph S. Kuykendall's belief that it was then that venereal disease was introduced into Hawaii.

Leaving the islands, Cook continued north for the next eight months, explored the western coasts of North America, and passed through Bering Strait into the Arctic Ocean. He did not, of course, find the Northwest Passage, and as the Arctic winter began to close in he turned back to Hawaii, because the *Resolution* was sadly in need of repairs. On the morning of Sunday, November 30, 1778, he sailed into Kealakekua Bay on the western side of Hawaii and dropped anchor. The ancient god Lono had at last returned.

Kalaniopuu, at that time chief of the district of Kau, made this Lono more than welcome. As Kuykendall relates, "Every day the Hawaiians sent aboard large quantities of hogs and vegetables. On January 25 [1779] Kalaniopuu again visited Cook, exchanged names with him, and presented him with several feather cloaks. Cook in return gave the king a linen shirt, a sword, and later a 'complete tool chest.' During the visit the natives entertained with boxing and wrestling matches, and the visitors gave a display of fireworks."

Such was the amicable prelude to the incredible display of human folly that was to follow. On February 14 the Hawaiians were irritable and out of patience with the "enormous consumption of hogs and vegetables." This irritability spread like a contagion to the men of the *Resolution* and *Adventure*. Armed, they put out in small boats for the shore. A Hawaiian chief was shot from a small boat. A marine was slain with a dagger. Cook himself fired two shots; the first was a blank, but the second killed a man. More marines began shooting, and before Cook could stop them, a score of natives had been killed. Cook then went ashore to try to calm the general hysteria, but a Hawaiian with a club crept up on him from behind and hit him on the back of the head. David Samwell, the surgeon, told what followed.

> The stroke seemed to have stunned Captain Cook: he staggered a few paces, and fell on his hands and one knee, and dropped his musket. As he was rising, and before he could recover his feet, another [Hawaiian] stabbed him in the back of the neck with an iron dagger. He then fell into a bite of water about knee deep, where others crowded upon him, and endeavoured to keep him under; but struggling very strongly with them, he got his head up, and casting his look towards the pinnace, seemed to solicit assistance. Though the boat was not above five or six yards distant from him, yet from the crowded and confused state of the crew, it seems, it was not in their power to save him. The [Hawaiians] got him under again, but in deeper water; he was, however, able to get his head up once more, and being almost spent in the struggle, he naturally turned to the rock, and was endeavouring to support himself by it, when a savage gave him a blow with a club, and he was seen alive no more....
>
> I need make no reflection on the great loss we suffered on this occasion, or attempt to describe what we felt. It is enough to say that no man was ever more beloved or admired; and it is truly painful to reflect, that he seems to have fallen a sacrifice merely for want of being properly supported; a fate singularly to be lamented, as having fallen to his lot, who had ever been conspicuous for his care of those under his command, and who seemed, to the last, to pay as much attention to their preservation, as to that of his own life.

Cook was a few months over fifty years of age.

When the news of Cook's death reached England, King George wept, and one who signed himself "Columbus" wrote a moving eulogy for the *Morning Chronicle*. Observers believed that "Columbus" was Sir Joseph Banks—who had accompanied Cook on the first voyage and who was for many years president of the Royal Society.

Samwell himself wrote: "The character of Captain Cook will be best exemplified by the services he has performed, which are universally known, and which have ranked his name above that of any navigator of ancient or of modern times. Nature had endowed him with a mind vigorous and comprehensive, which in his riper years he had cultivated with care and industry. His general knowledge was extensive and various; in that of his profession he was unequalled. With a clear judgment, strong masculine sense, and the most determined resolution; with a genius particularly tuned for enterprise, he pursued his object with unshaken perseverance:—vigilant and active in an eminent degree: cool and intrepid among dangers: patient and firm under difficulties and distress: fertile in expedients: great and original in all his designs, active and resolved in carrying them into execution. These qualities rendered him the animating spirit of the expedition; in every situation he stood unrivalled and alone: on him all eyes were

This monument at Kealakekua Bay marks the spot where Captain Cook died. Coral reefs guard the inlet from the open sea, and ancient Hawaiian burial caves dot the surrounding cliffs.

turned: he was our leading star, which at its setting left us involved in darkness and despair."

The world could well share David Samwell's sense of loss. Ranked in science with Sir Isaac Newton and Charles Darwin, Cook was in temperament and pace much more like Darwin. Sir Isaac had made all his important discoveries by the time he was twenty-four. Darwin continued a full productive life until his death at seventy-three. Had Cook been spared another decade or more, his world might well have been a wiser place.

ately upon my settling in Birmingham. He encouraged me in the prosecution of my philosophical enquiries, and allowed me forty pounds per annum for expenses of that kind, and was pleased to see me make experiments to entertain his guests, and especially foreigners."

Priestley called these volumes "Experiments and Observations on Different Kinds of Air," and it is important to note that all of the substances that he called "airs" were actually gases. Thus carbon dioxide in Priestley's terms was "fixed air," ammonia was "acid air," oxygen was "dephlogisticated air," common air lacking in oxygen was "vitiated air," and so on.

One of Priestley's early interests was in the "fixed air" that blankets the vats in which beer is brewed, and soon he had the idea of dissolving this carbon dioxide in water under pressure. Hence, of course, soda water, which took Europe by storm.

But of all of his investigations on "air," Priestley's most important contribution had its roots in an experiment that he did as a boy. According to his brother Timothy, "My brother began to discover a taste for experiments when he was about eleven years of age. The first he made was on spiders, and by putting them into bottles, he found how long they could live without fresh air."

In the most momentous experiments of his mature years, Priestley found that if he put a green plant into a closed vessel of air from which the oxygen had been consumed by a burning candle, the plant would restore the oxygen. In his report to the Royal Society of London, he said:

"I flatter myself that I have accidentally hit upon a method of restoring air which has been injured by the burning of candles, and that I have discovered at least one of the restoratives which nature employs for this purpose. It is vegetation. In what manner this procedure in nature operates, to produce so remarkable an effect I do not pretend to have discovered; but a number of facts declare in favour of this hypothesis. I shall introduce my account of them, by reciting some of the observations which I made on the growing of plants in confined air, which led to this discovery."

In this day of elaborate and costly laboratory apparatus, it is interesting to note the simplicity of the equipment that Priestley used. He confined his gases in upturned glass jars sealed off at the mouth by inverting them in water. Some of these were simple open-mouth cylinders that he had used for his electrical experiments, and others were conical beer glasses. As containers for the water he used tea dishes, or, for a larger reservoir of water,

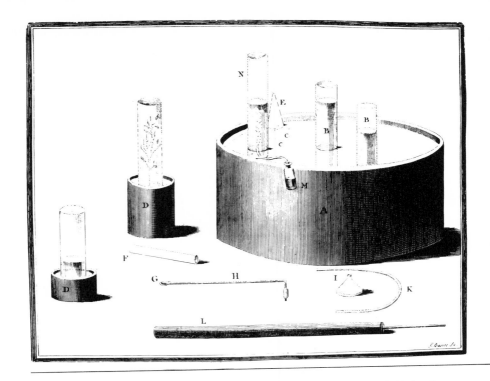

With the simple apparatus depicted in this old cut, Joseph Priestley showed that green plants produce oxygen. His own description of the apparatus was as follows: "The apparatus with which the principal of the preceding experiments were made is exceedingly simple, and cheap. The drawing annexed exhibits a view of every thing that is most important in it. A is an oblong trough, about eight inches deep, kept nearly full of water, and B,B are jars standing in it, about ten inches long, and two and a half wide; such as I have generally used for electrical batteries. C,C are flat stones, sunk about an inch, or half an inch, under the water, on which vessels of any kind may be conveniently placed, during a course of experiments. D,D are pots nearly full of water, in which jars or phials, containing any kind of air, to which plants or any other substances may be exposed, and having their mouths immersed in water; so that the air in the inside can have no communication with the external air. E is a small glass vessel, of a convenient size for putting a mouse into it, in order to try the wholesomeness of any kind of air that it may contain. F is a cylindrical glass vessel, five inches in length, and one in diameter, very proper for trying whether any kind of air will admit a candle to burn in it. For this purpose a bit of wax candle, G, may be fastened to the end of a wire, H, and turned up in such a manner to be let down into the vessel with the flame upwards. The vessel should be kept carefully covered till the moment that the candle is admitted to it. In this manner I have frequently extinguished a candle above twenty times in one of these

an oval stoneware trough that was commonly used for washing linen.

"One might have imagined," Priestley continues, "that since common air is necessary to vegetable, as well as to animal life, both plants and animals had affected it in the same manner, and I own that I had that expectation, when I first put a sprig of mint into a glass jar standing inverted in a vessel of water; but when it had continued growing there for some months, I found that the air would neither extinguish a candle nor was it at all inconvenient to a mouse, which I put into it."

Priestley found further that his plants themselves were not affected by his experimental conditions any more than one might expect from the small size of the jars in which he kept them. As time passed, the new sets of leaves grew smaller and smaller, however, and eventually the stems rotted. Anyone who wished to repeat his experiments, he pointed out, should be careful to keep the space clear of dead leaves.

Having found that candles burned well in air in which plants had been grown for some time, Priestley's next step was to guess

---

vessels full of air, though it is impossible to dip the candle into it, without giving the external air an opportunity of mixing with it, more or less. I is a funnel of glass or tin, which is necessary for transferring air into vessels which have narrow mouths. K is a glass syphon, which is very useful for drawing air out of a vessel which has its mouth immersed in water, and thereby raising the water to whatever height may be most convenient. I do not think it by any means safe to depend upon a valve at the top of the vessel, which Dr. Hales very often made use of; for, since my first disappointment, I have never thought the communication between the external and internal air sufficiently cut off, unless glass, or a body of water, or, in some cases, quicksilver, have intervened between them. L is a piece of a gun barrel, closed at one end, having the stem of a tobacco-pipe luted to the other. To the end of this pipe I sometimes fastened a flaccid bladder, in order to receive the air discharged from the substance contained in the barrel; but, when the air was generated slowly, I commonly contrived to put this end of the pipe under a vessel full of water, that the new air might have a more perfect separation from the external air than a bladder could make. M is a small phial containing some mixture that will generate air. This air passes through a bent glass tube inserted into the cork at one end, and going under the edge of the jar N at the other; the jar being placed with part of its mouth projecting beyond the flat stones C,C for that purpose.

that plants might also restore air that had been "injured" by burning a candle in it. To test this, he "injured" the air in a glass cylinder by burning a wax candle in it until the flame died. Then he put a sprig of mint into the container and left it for ten days. At the end of that time the candle burned perfectly well in the "restored" air. He repeated the experiment many times during that summer and always found the same result.

As he varied his experimental conditions, he found that "injured" air could be "restored" by sprigs of mint in five or six days, while no other treatment that he could devise (such as rarefying the air, or heating it, or exposing it to light) had any effect.

A year later he repeated the experiments, and this time, in addition to burning candles to "injure" the air, he burned brandy and sulfur matches. In these experiments also, mint plants could restore the "injured" air. Nor was mint the only plant that did so. Sprigs of balm did just as well, as did groundsel, a weed, and spinach.

We have known since the time of Lavoisier, of course, that Priestley's air was "injured" through the depletion of its oxygen, either by flames or by animal respiration; and we shall shortly see the explanation of just how plants "restored" it. In his own lifetime, however, Priestley never accepted Lavoisier's revolutionary chemistry—an ironic twist in the reasoning of a man whose very experiments were in perfect accord with Lavoisier's conclusions. Not many years before Priestley died, and well after Lavoisier's chemistry was generally accepted, Priestley could still write, ". . . I must observe that, in the presentation of my experiments, I was led to maintain the doctrine of Phlogiston[5] against Mr. Lavoisier, and other chemists in France, whose opinions were adopted not only by almost all the philosophers of that country, but by those in England and Scotland also." It is a strange anomaly that such a great experimental chemist should have so often drawn the wrong conclusions from his own excellent work. Of him the great French naturalist Baron Georges Cuvier was moved to remark, "He was the father of modern chemistry, but he would never acknowledge his child."

On the other hand, it is encouraging for the human condition that a man who could be so wrong so often should make so many

---

[5] Before Lavoisier, all combustible substances were supposed to be made up of phlogiston and ash; when a substance was burned, phlogiston was supposed to escape, and conversely, the original substance could be produced by adding phlogiston to the ash.

wonderful discoveries. He prepared oxygen by heating mercuric oxide, and discovered ammonia, sulfur dioxide, carbon monoxide, hydrochloric acid, nitric oxide, and hydrogen sulfide. As Sir Humphry Davy said, "No single person ever discovered so many strange and curious substances."

It would be hard to imagine two great scientists whose backgrounds and personalities differed more sharply than did those of Joseph Priestley and Jan Ingenhousz. Priestly, "champion of freedom," was anti-Establishment throughout his life, and his outspoken nonconformist religious beliefs finally led unruly mobs to burn down his chapel in Birmingham and his house as well. When he continued his religious harangues in London, public feeling ran so strongly against him that he felt constrained to emigrate to America. He left England in 1794 and spent the last ten years of his life in Northumberland, Pennsylvania.

Jan Ingenhousz, on the other hand, was a model of conformity, a consequence, perhaps, of his warmest of dispositions. Lord Shelburne, Priestley's benefactor, once said that he always believed that Jeremy Bentham (the advocate of "the greatest good for the greatest number") was the most good-natured man in the world—until he met Ingenhousz.

Jan Ingenhousz was born in 1730 in Breda, the Netherlands, to a prominent local family. His father, Arnold Ingenhousz, was a successful businessman, and young Jan had the finest of educations. He studied medicine in Breda and in Louvain, receiving his M.D. at the age of twenty-two, and then continued studies both in medicine and in science in Leiden, Paris, and Edinburgh. It was during this period that he became fascinated with the idea of doing original research in science, and it was with this in mind that he returned to Breda in 1757. But the medical practice that he established became so successful that his plans for wider scientific accomplishments began to fade. They were not to reach full fruition for another twenty years.

Ingenhousz continued his medical practice in Breda for some ten years. During this period, Sir John Pringle, the London authority on military sanitation, often visited the Ingenhousz family in Breda, and was deeply impressed with the ability of young Dr. Jan. When Jan's father died in 1766, he moved to London and continued his medical practice under Sir John's influential sponsorship. During these London days, Ingenhousz developed his highly successful procedure for inoculation against smallpox.⁶

At just that time Vienna was ravaged by an epidemic of the scourge, and Empress Maria Theresa asked for the best man in England to inoculate her surviving children and also to serve as surgeon to the royal family. Sir John Pringle recommended Ingenhousz, and in 1768, with the approval of King George III, he went to the Viennese royal court. Daydreams of original investigations in science still drifted through his mind, and from time to time he even dabbled with some experiments on plants. But his main obligation (and attraction) was to the royal court, with all the opulence and glitter of eighteenth-century Vienna.

Although we do not find explicit mention of it in Ingenhousz's own writings, or in those of his biographers, it seems quite possible that the turning point in Ingenhousz's scientific life came with his marriage to Agathe Maria Jacquin. Ingenhousz was then forty-five, and his bride's father was the well-known professor of botany and chemistry at the University of Vienna, Nikolaus Jacquin. Perhaps this renewed connection with the academic environment may have modified Ingenhousz's commitment to court society.

In any event, he asked the Empress for a three-month leave of absence in the summer of 1778, and, returning to England to work, he performed a vast number of experiments on the mechanism of absorption and evolution of gases by plants. It was indeed an impressive summer's work for a man long used to the ease of Maria Theresa's court, with such luxuries as its dinners of "fifty dishes of meat, all served in silver, and well dressed ... and [wines] to the number of eighteen different sorts, all exquisite in their kinds."

Of that summer Ingenhousz wrote, in *Experiments upon Vegetables,* published in 1779, "On purpose to avoid every cause of obstructing my mind in pursuit of the object I had in view, and in tracing Nature in its operation on this subject, I disengaged myself from the noise of the metropolis, and retired to a small villa, where I was out of the way of being interrupted by anybody in the contemplation of Nature.

"This work is a part of the result of above 500 experiments, all of which were made in less than three months, having begun

---

⁶ Although this procedure was more hazardous than the vaccination technique that Edward Jenner developed thirty years later, it nevertheless saved many lives.

them in June, and finished them in the beginning of September, working from morning till night. From these experiments some more consequences might have been drawn, if I had had more time to employ myself in a work upon such important matter. Whatever I have been able to deduce from my labours is done in a hasty manner, as my stay in this country was far too limited to allow me to compose my work in a regular and more satisfactory manner.

"Though I was far from foreseeing all the discoveries which I made in the course of this summer, yet I was persuaded that a good deal of the economy of the vegetable kingdom might be discovered by a steady pursuit of experiments tending to trace the operations of Nature. I had this object in view some years ago; but as I did not enjoy such a favourable disposition of mind and body as was necessary for a task, in which all possible steadiness, perseverance, and close attention were requisite, I deferred the undertaking till I should find myself fit for it."

Ingenhousz returned briefly to Maria Theresa's court, but in 1779 he moved permanently to England, where he published the results and conclusions of his work of the previous summer. He called his book *Experiments upon Vegetables, Discovering Their Great Power of Purifying the Common Air in the Sun-Shine, and of Injuring It in the Shade and at Night.*

As with Priestley's, Ingenhousz's work was done before Lavoisier's explanations of respiration and combustion were generally accepted, and for the convenience of today's reader, I have updated the outmoded terms that Ingenhousz used in that edition of his work.

"I was not long engaged in this enquiry before I saw a most important scene opened to my view; I observed that plants not only have a faculty to correct bad air in six to ten days, by growing in it, as the experiments of Dr. Priestley indicate, but that they perform this important office in a complete manner in a few hours; that this wonderful operation is by no means owing to the growth of the plant, but to the influence of the light of the sun upon the plant. I have found that plants have, moreover, a most surprising faculty of elaborating the air which they contain, and undoubtedly absorb from the common atmosphere, into real and fine natural air; that they pour down continually, if I may so express myself, a shower of this purified air, which, diffusing itself through the common mass of the atmosphere contributes to render it more fit for animal life; that this operation is far from being carried on constantly, but begins only after the sun has for some time made its appearance above the horizon, and has, by its influence, prepared the plants to begin anew their beneficial operation upon the air, and thus upon the animal creation, which has stopped during the darkness of the night; that this operation of the plants is more or less brisk in proportion to the clearness of the day, and the exposure of the plants more or less adapted to receive the direct influence of that great luminary; that plants shaded by high buildings, or growing under the dark shade of other plants, do not perform this office, but on the contrary throw out an air hurtful to animals, and even contaminate the air which surrounds them; that this operation of plants diminishes toward the close of the day, and ceases entirely at sunset, except in a few plants, which continue this duty somewhat longer than others; that this office is not performed by the whole plant, but only by the leaves and the green stalks that support them; that acrid, ill-scented, or even the most poisonous plants perform this office in common with the mildest and most salutary; that leaves pour out

Microscopic marine diatoms come in a great variety of shapes. All of them contribute to the conversion of the sun's energy to primary food for the biomass of the sea.

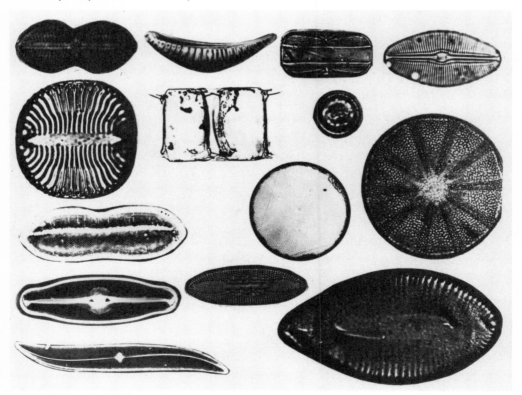

A marine diatom at 3,000× magnification. This marine diatom, Asteromphalus robustus, is just one of thousands of different species of diatoms that participate in the ocean's vast photosynthetic process. Photographed at 3,000× by the Scanning Electron Microscope at Scripps Institution of Oceanography.

the greatest quantity of oxygen from their under surface, principally those of lofty trees; that young leaves, not yet come to their full perfection yield less oxygen than is produced by full-grown and old leaves; that some plants produce oxygen better than others; that some of the aquatic plants seem to excel in this operation; that all plants give off carbon dioxide at night, and even in the daytime in shaded places; . . . and that the sun by itself has no power to mend air without the concurrence of plants, but on the contrary is apt to contaminate it."

This, then, is the heart of the work of which Charles Singer said, "This discovery is the foundation of our whole conception of the economy of the world of living things."

As we saw, Ingenhousz's 1779 publication came before the general acceptance of Lavoisier's work on combustion and respiration, and was written against the background of the phlogiston doctrine. However, Ingenhousz continued his work in plant physiology for another twenty years, and toward the end of that time he wrote a new account of his work in the light of Lavoisier's sound chemistry. This was his *Essay on the Food of Plants and the Renovation of Soils,* which he published in 1796. By then it had become clear to him that plants utilize carbon dioxide in their nutrition and give off oxygen in the process.

The dawn of the nineteenth century brought a new series of brilliant investigations of the photosynthetic process. De Saussure made accurate measurements of the amounts of carbon dioxide that plants use under different intensities of light, as well as the amounts of oxygen that they produce. Daubeny showed that chlorophyll is necessary for photosynthesis. Sachs showed that plants produce starch during photosynthesis. It then became possible to write the basic chemical equation for what takes place when a plant transforms the sun's radiant energy into stored chemical energy in the form of a carbohydrate:

$$CO_2 + H_2O \xrightarrow[\text{green plant}]{\text{light energy}} (CH_2O) + O_2$$

Of course, green plants all around us on land and sea are storing energy in this way throughout the sunny hours. Of particular concern to us is the proportion of it that takes place in the world's oceans. Isaac Asimov has summarized this very well:

"The scale on which the earth's green plants manufacture organic matter and release oxygen is enormous. The Russian-born American biochemist Eugene I. Rabinowitch, a leading investigator of photosynthesis, estimates that each year the green plants of the earth combine a total of 150 billion tons of carbon (from carbon dioxide) with 25 billion tons of hydrogen (from water) and liberate 400 billion tons of oxygen. Of this gigantic performance, the plants of the forests and fields account for only 10 per cent; for 90 per cent we have to thank the one-celled plants and seaweed of the oceans."

5.
FORBES: THE DAWN OF MARINE ECOLOGY

Today, more than a century after he died, Edward Forbes, the Manx naturalist, is acclaimed as the outstanding pioneer in the ecology of the seas—a new science. It is ironic that the term "ecology" had not even been invented in Forbes' day, and that he could not have imagined the richness of the field that his work was to open up.

Now, with the clarity of 20/20 hindsight, we can appreciate the truly innovative nature of Forbes' work, for we can compare what he did and wrote with the ideas of ecology as they have developed since. Let us therefore move ahead of our story for a moment to some concepts of ecology that emerged well after Forbes' time.

One of the earliest definitions of ecology was given by Ernst Haeckel, of the University of Vienna, in 1870—some thirty years after Edward Forbes had given a clear statement about population dynamics, one of the most abstruse and difficult fields in all of present-day ecology. Haeckel was a prolific writer, both for the German scientific community as well as for the lay public, urging

Darwin's views on everyone who would listen. Of ecology, he said, "By ecology we mean the body of knowledge concerning the economy of nature—the investigation of the total relations of the animal both to its inorganic and to its organic environment; including, above all, its compatible and incompatible relations with those animals and plants with which it comes directly or indirectly into contact. In a word, ecology is the study of all those complex interrelations referred to by Darwin as the conditions of the struggle for existence. The science of ecology . . . has thus formed the principal component of what is commonly referred to as 'natural history.' As is well shown by the numerous popular histories of both early and modern times, this subject has developed in the most close relations with systematic zoology."

In the sixty years that followed Haeckel's nineteenth-century definition, the content and method of ecology have been much clarified, and by 1933 Charles Elton could write: "Ecology represents partly the application of scientific method in natural history, and partly something more. . . . When the species of animals [that have been observed by naturalists] are properly identified by an agreed system of names, specimens stored in museums for reference, the habitats of the animals accurately described, the information tested by various different workers, ingenious experiments devised to find out causes, then we can say that natural history has begun to be animal ecology. Ecology means the relation of animals and plants to their surroundings. As a science it may be said to depend on three methods of approach: field observations, adequate systematic technique for determining the names of the animals, and experimental work both in the field and in the laboratory."

Although it was first published in 1933, Elton's *The Ecology of Animals,* from which I have just quoted, has survived three editions and many reprintings and is still available. Readers who may have been annoyed and confused by hearing the term "ecology" applied to everything from the collection of empty beer cans to the design of nuclear power plants will find it excellent reading.

Let us return to Edward Forbes in the early decades of the nineteenth century. What led him to his interest in the sea and in the creatures within its waters? He was a highly articulate man, and his own words tell a convincing story.

> The investigation and determination of provinces of marine life have as yet been little pursued, and there is no

finer field for discovery in natural history, than that presented by the bed of the ocean, when examined with a view to determining its natural subdivisions. The difficulties which attend the inquiry add to the zest of the research; and there is a charm in travelling mentally over the hills and valleys buried inaccessibly beneath their thick atmosphere of brine, unbreathable by mortal lungs, which air travelling, being an easy possibility, and its results, do not possess. . . .

To sit down by the sea-side at the commencement of the ebb, and watch the shore gradually uncovered by the retiring waters, is as if a great sheet of hieroglyphics—strange picture-writing—were being unfolded before us. Each line of the rock and strand has its peculiar characters inscribed upon it in living figures, and each figure is a mystery, which, though we may describe the appearance in precise and formal terms, has a meaning in its life and being beyond the wisdom of man to unravel. How many and how curious problems concern the commonest of the sea-snails creeping over the wet seaweed! In how many points of view may its history be considered! There are its origin and development—the mystery of its generation—the phenomena of its growth—all concerning each apparently insignificant individual; there is the history of the species—the value of its distinctive marks—the features that link it with higher and lower creatures—the reason why it takes its stand where we place it in the scale of creation—the course of its distribution—the causes of its diffusion—its antiquity or novelty—the mystery (deepest of mysteries) of its first appearance—the changes of the outline of continents and of oceans which have taken place since its advent, and their influence on its own wanderings. Some of these questions may be clearly and fairly solved; some of them may be theoretically or hypothetically accounted for; some are beyond the subtlety of human intellect to unriddle. I cannot revolve in my mind the many queries which the consideration of the most insignificant of organized creatures, whether animal or vegetable, suggests, without feeling that the rejection of a mystery, because it is a mystery, is the most besotted form of human pride.

Edward Forbes was born at Douglas, on the Isle of Man, in 1815. His father was a successful banker, and his mother was the heiress of the ancestral estates of Corvalla and Ballabeg; this

This ascidian, Diazona violacea, *was found by Edward Forbes in the Hebrides in 1850. Known to Forbes as* Syntethys hebridica, *it is green when alive and violet when dead. About 65 per cent of natural size.*

family competence may account in part for young Edward's indifferent performance as a student. Nevertheless, he could apply himself with great energy to those projects in which he took a real interest. And surely he had ample self-confidence with respect to these. At the age of twelve he completed a manuscript that he entitled *A Manual of British Natural History in All Its Departments.*

In those days, of course, only three careers were open to a young man of good birth: the church, the law, and medicine. Any other calling would have been quite unthinkable. Young Edward's mother favored the church; his father favored medicine; and Edward avoided filial disapproval either way by going to London to study art. No more than four months of that were needed to convince both teacher and pupil alike that brush and canvas were not for Forbes (although he did do quite well at sketching with a pencil). Medicine, then, seemed somewhat related to the world of nature that continued to fascinate young Forbes, and he entered the University of Edinburgh as a medical student in the fall of 1831.

His student notebooks give ample evidence of his lack of interest in these studies. He would start his notes for a lecture period with a few lines of serious summary. These would gradually deteriorate into incoherence. After a pause they would begin again, now in the vein of satirical comment on the personality of the lecturer. Often the day's notes would end with a pencil sketch of the lecturer, usually in caricature, and usually signed with the monogram "BBBB"—that is, four "B's"—the Scottish pronunciation of the name Forbes.

There is a suggestion of this same impish nature in one well-known portrait. Forbes' pose is apparently that of a dignified man in correct attire, his right forearm held formally across his waist. At first glance his gaze seems serious indeed, but the closer one studies the face, the more likely it seems that the subject is struggling to contain a grin.

There is more than a suggestion of this nature in Forbes' organization at Edinburgh of the Brotherhood of the Magi, a select dining club that was devoted to "good fellowship, good talk, and good literature." They did not need to add: "good wine." Each meeting of the Magi was opened by nine standing toasts, which may have made more bearable some of the doggerel that was perpetrated, especially by Forbes.

After four long years of such medical "study," the university authorities called Forbes to face a comprehensive examination. He had yawned over bones in the anatomy laboratory, filled his lecture notes with doggerel and lampoons of his professors, and spent many diverting evenings with the Brotherhood of the Magi. What then of exposing the meagerness of his medical learning before a battery of dour Edinburgh professors? Forbes's response was simple and direct. He just did not appear. His record says: *"Non inventus."*

Why, we ask, should Edward Forbes have pursued this open sham of studying medicine for so long? We do not know. Only one full-length biography of him has been written, and that was by men whose preoccupation with science itself left little room for any inquiry into human motivation. The *Memoir* of Wilson and Geikie is quite dry going. But whatever Forbes' reasons, the end of these misdirected farces lifted a great cloud from his mind. Free at last to throw himself into the study of natural history, he did it with all his might, and for the next five years he reveled in the joy and stimulation of that work.

He spent his summers in the field, studying animals and

plants in their natural environments and collecting specimens. He explored the fauna and flora of his native Isle of Man. He backpacked into Norway, and while in Bergen he collected a sample of fine shell sand from a spittoon, for later study under the microscope. Several summers he spent dredging the Irish Sea for marine specimens. He was not the first to use the dredge for this purpose, but his results were some of the most fruitful.

In 1839 he and John Goodwin dredged the Shetland Sea, and that summer Forbes presented the results of the work to the British Association for the Advancement of Science. The findings were so enthusiastically received that a Dredging Committee was set up forthwith, and official support of such work was established on a continuing basis. Even with such official acclaim, however, Forbes did not take either his work or himself too seriously, and he was soon one of the founding members of another dining and good-fellowship group—this time the Red Lion Clubbe. For its first meeting he presented to the membership his "Song of the Dredge," which went in part as follows.

> *Hurrah for the dredge, with its iron edge,*
> > *And its mystical triangle.*
> *And its hided net, with meshes set*
> > *Odd fishes to entangle!*
> *The ship may move thro' the waves above,*
> > *'Mid scenes exciting wonder,*
> *But braver sights the dredge delights*
> > *As it roves the waters under.*
>
> CHORUS: *Then a-dredging we will go, wise boys*
> > *A-dredging we will go!*
> *A-dredging we will go, a-dredging we will go,*
> > *A-dredging we will go, wise boys, wise boys,*
> *A-dredging we will go.*
>
> *Down in the deep, where the mermen sleep,*
> > *Our gallant dredge is sinking;*
> *Each finny shape in a precious scrape*
> > *Will find itself in a twinkling!*
> *They may twirl and twist, and writhe as they wist,*
> > *And break themselves into sections,*
> *But up they all, at the dredge's call,*
> > *Must come to fill collections.*

Then a-dredging, etc.

The creatures strange the sea that range,
　　Though mighty in their stations,
To the dredge must yield the briny field
　　Of their loves and depredations.
The crab so bold, like a knight of old,
　　In scaly armour plated,
And the slimy snail with a shell on his tail,
　　And the star-fish—radiated!

Then a-dredging, etc.

By now Forbes was well into the work that was to continue for the rest of his life, and which involved all of the major divisions that we now think of as making up the study of ecology. Briefly they are as follows.

• Surveys to determine the kinds of animals and plants that live in various environments.
• Studies of interrelations between the various species in an environment.

Forbes' lampoon of the naturalists' dredge.

• Definitive descriptions of typical habitats.
• Statistical studies to determine animal populations at specified times.
• Statistical studies to determine the variations in populations from one time to another.
• Economic problems involved in man's use of these natural resources.

Long before they had been so defined, Edward Forbes was carrying out studies along most of these lines. A few examples will serve to make the point.

As late as 1933, Charles Elton wrote, "The attractions are that ecological surveys bring the student into direct contact with live animals in a state of nature, while at the same time almost always leading to special problems whose existence would not otherwise have been suspected."

A century earlier, in his *History of British Starfishes,* Edward Forbes had written, "The first time I ever took one of these creatures [the starfish *Luidia fragilissima*] I succeeded in getting it into the boat entire. Never having seen one before, and quite unconscious of its suicidal powers, I spread it out on a rowing bench, the better to admire its form and colours. On attempting to remove it for preservation, to my horror and disappointment, I found only an assemblage of rejected members. My conservative endeavours were all neutralized by its destructive exertions, and it is now represented in my cabinet by an armless disk and a diskless arm. Next time I went to dredge on the same spot, determined not to be cheated out of a specimen in such a way a second time, I brought with me a bucket of cold fresh water, to which article Starfishes have a great antipathy. As I expected, a *Luidia* came up in the dredge, a most gorgeous specimen. As it does not generally break up before it is raised above the surface of the sea, cautiously and anxiously I sank my bucket to a level with the dredge's mouth, and proceeded in the most gentle manner to introduce *Luidia* to the purer element. Whether the cold air was too much for him, or the sight of the bucket too terrific, I know not, but in a moment he proceeded to dissolve his corporation, and at every mesh of the dredge his fragments were seen escaping. In despair I grasped the largest, and brought up the extremity of an arm with its terminating eye, the spinous eyelid of which opened and closed with something exceedingly like a wink of derision."

In studying the distribution of marine animals, Forbes ob-

The starfish Luidia ciliaris, *a close relative of* Luidia fragilissima, *which Forbes could collect only as "an armless disk and a diskless arm."*

served relationships with climate and with temperature, with depth as affecting both pressure and available light, with the nature of the sea bottom, with the effects of tides and currents, and with the composition of the sea water.

Concerning the salinity of the water he wrote, "In many confined localities, as in the lochs of Scotland, and the fiords of Norway, also in many estuaries, the surface waters may be fresh, or nearly so, whilst their depths are as salt as the open ocean, so that in the same place we may have creatures organized for very different states of sea-composition living not merely in the immediate neighborhood of each other, but even, as it were, superimposed. I was once greatly struck with this fact, when dredging in the Killeries, along with Robert Ball and William Thompson, an arm of the sea in the wild and rocky district of Connemara, in Ireland. The depth was some fifteen or twenty fathoms, and the creatures inhabiting the sea-bottom were characteristically marine. When taken out of the water, they seemed to be unusually torpid, and it was in vain we placed them in vessels filled with the element of their native bay in order to tempt them to display their variously-shaped delicate organs. The cause of their languor soon became evident when we remarked a fisherman dipping a cup into the water by the boat-side for the purpose of procuring some to drink. The uppermost stratum of the narrow and lake-like bay was purely fresh, or nearly so, derived doubtless from the numerous streamlets flowing into it, and from the rain, over-sufficiently abundant in that mountainous and picturesque district. The mollusca and radiata drawn from the salt waters beneath, became convulsed and paralysed in their involuntary ascent through the fresh waters above. They were more dead than alive when we placed them in basins, and none the livelier for having a new supply of water taken from the surface of the sea. Yet whilst these truly marine creatures were living and thriving below, numerous forms of entomostraca, incapable of enduring the briny fluids of the depths, might be sporting in the lighter and purer element above."

Forbes was also fully aware of the influence of the depth of the sea upon the kinds of plants and animals to be found in successive regions or zones. From the highest of high-water marks to the greatest depths, he delineated five zones, which he called the *littoral,* the *laminarian,* the *coralline,* the *deep-sea coral;* and finally "an abyss where life is either extinguished, or exhibits but a few sparks to mark its lingering presence." Each of these zones had its own characteristic distribution of plants and animals, with respect to both the numbers of different species found and the numbers of individuals in each species.

The richly populated *littoral* zone was that between the tide marks, and contained plants and animals that can withstand (or perhaps need) periodic exposure to the air, to direct light, and to wind and rain. The *laminarian* zone extended from the edge of low water to a depth of about fifteen fathoms. Forbes found it to abound in seaweeds and to be exceptionally rich in fishes, mollusks, crustaceans, and innumerable invertebrates. "This region, above all others, swarms with life, and when we look down through the clear waters into the waving forests of broad-leaved tangles, we see animals of every possible tint sporting among their foliage, darting from frond to frond, prowling among their gnarled roots, or crawling with slimy trail along their polished bronze expanses."

Below the *laminarian,* Forbes' *coralline* zone extended from fifteen fathoms to thirty or more, and contained many coral-like nullipores, as well as plants that looked like minerals. And farther down, into the zone of the *deep-sea corals,* the living forms were few, comprising mainly animals that looked more like plants and had stony shells. Below this, Forbes expected to find no living forms except some that might have strayed there.

All these studies were carried out in the British seas. But in 1841, Forbes was offered the post of naturalist aboard H.M. Surveying Ship *Beacon,* which was doing hydrographic work in the eastern Mediterranean. Here was a golden opportunity to extend his studies into other waters, and Forbes eagerly accepted. The nature of the expedition gave him time and opportunity to make a wide-ranging series of studies on the statistical distribution, at different depths, of many species of animals in the Aegean Sea. One of the most remarkable of these was his collection of data on the depth distribution of shellfish. The results that he found, and later presented to the British Association for the Advancement of Science, are shown in the accompanying table.

NUMBERS OF SPECIES OF SHELLFISH IN THE AEGEAN SEA FOUND BY EDWARD FORBES AT DIFFERENT DEPTHS, 1841

PHYLUM	INTERTIDAL	DEPTH IN FATHOMS							
		2	5	20	35	55	79	105	230
Chitons (sea cradles)	7	3	2	0	2	2	1	1	0
Patelliform univalves (limpets)	20	11	3	2	3	5	6	6	1
Dentalia (tooth shells)	6	4	4	2	2	1	1	2	2
Spiral univalves, holostomatous	115	50	40	40	44	35	28	17	15
Spiral univalves, phonostomatous	104	40	27	30	41	36	30	16	5
Pteropods (sea butterflies) and Nucleobranches	12	1	0	0	0	0	0	3	12
Brachipods (lamp shells)	8	0	0	0	2	4	5	7	3
Lamellibranches (certain mollusks)	135	38	53	52	68	58	48	34	28
AEGEAN TOTAL	407	147	129	126	162	141	119	86	66

Of all of today's divisions of ecology, "population dynamics"—the statistics of the variation of populations with time—is the most advanced and difficult. But even in this area, Forbes made some penetrating speculations from his observations in the Aegean Sea.

"As each region shallows or deepens," he told the British Association for the Advancement of Science, "its animal inhabitants must vary in specific association, for the depression which may cause one species to dwindle away and die will cause another to multiply. The animals themselves, too, by their over-multiplication, appear to be the cause of their own specific destruction. As the influence of the nature of the sea bottom determines in a great measure the species present on that bottom, the multiplication of individuals dependent on the rapid reproduction of successive generations of Mollusca will of itself change the ground and render it unfit for the continuation of life in that locality until a new layer of sedimentary matter, uncharged with living organic contents, deposited on the bed formed by the exuviae of the exhausted species, forms a fresh soil for similar or other animals to thrive, attain their maximum, and from the same cause die off."

Forbes was also deeply interested in the wise management of the resources of the sea. For example, in 1847 he wrote to a friend: "On Friday night I lectured at the Royal Institution. The subject was the bearing of marine researches and distribution matters on the fisheries question. I pitched into government mismanagement pretty strong, and made a fair case of it. It seems to me that at a time when half the country is starving we are utterly neglecting or grossly mismanaging great sources of wealth and food. . . . Were I a rich man, I would make the subject a hobby, for the good of the country and for the better proving that the true interests of government are those linked with and inseparable from Science."

It remains to note the mistake for which Forbes is so often cited. It seems incredible that a man who observed so keenly, and who reasoned so well in most things, could have made it. But he did. When he found that the deeper he went in the ocean, the fewer forms he recovered (and that below two hundred and thirty fathoms there were almost none), he reasoned that below three hundred fathoms life could not exist at all. No one in his own day could dispute it, but, as we shall see in a later chapter, life has been found in profusion at depths much greater than three hundred fathoms. Perhaps the very clarity of Forbes's statement was what led others to pursue the question; as Aristotle said: "Truth emerges more readily from error than from confusion."

In this account I have intentionally stressed Forbes' accomplishments in marine biology, for that is the focus of this book. However, he was also a fine geologist and paleontologist, and he held important posts in those fields, both as a means of livelihood and as an officer of the British Association for the Advancement of Science. But in spite of these wide interests, Forbes' lifelong ambition was to achieve appointment to the Chair of Natural History at the University of Edinburgh. This he finally realized in March 1854.

He was not to hold it for long. During the summer of 1854 he gave a course of lectures before a "large and enthusiastic audience," but by August he fell seriously ill and was forced to take a rest in the country. He returned to Edinburgh in the fall to prepare for his winter course, and managed to meet his appointments for a week, although he was obviously weakening rapidly. He ran a high fever, a severe cold added its toll to the kidney ailment that he had suffered for some months, and in ten days he died. He was not quite forty years old.

During this sadly brief lifetime Edward Forbes made outstanding contributions to the systematic description of life forms in British, Scandinavian, and Mediterranean waters. His many monographs on jellyfish, starfishes, snails, sea urchins, and mollusks are classics. He made extensive contributions to our knowledge of the distribution of marine life geographically, by zones of depth, and in relation to fossil records. Throughout all this he was in constant search for generalizations from which he might see some central logic relating apparently random facts. His success at making such generalizations is perhaps the outstanding feature of his work.

He was, in Sir William Herdman's view, "the most original, brilliant, and inspiring naturalist of his day, with a broad outlook over nature and a capacity for investigating the borderline problems involving several branches of science; he was, in a word, a pioneer of oceanography."

As if that were not enough, Forbes' work was the direct inspiration for the greatest oceanographic expedition ever undertaken—the voyage of the *Challenger* in 1872. As this "noble ship . . . equipped for scientific research as no ship of any nation was ever equipped before" lay in readiness off Sheerness, Charles Wyville Thomson, Regius Professor of Natural History at the University of Edinburgh and Director of the Civilian Scientific Staff of the *Challenger* Exploring Expedition, wrote:

". . . although we are now inclined to look somewhat differently on certain very fundamental points, . . . to Forbes is due the credit for having been the first to treat these questions in a broad philosophical sense, and to point out that the only means of acquiring a true knowledge of the rationale of the distribution of our present fauna, is to make ourselves acquainted with its history, to connect the present with the past. This is the direction which must be taken by future inquiry. Forbes, as a pioneer in this line of research, was scarcely in a position to appreciate the full value of his work. Every year adds enormously to our stock of data, and every new fact indicates more clearly the brilliant results which are to be obtained by following his methods, and by emulating his enthusiasm and indefatigable industry."

The fifty volumes of the *Challenger* Reports are thus a fitting tribute to the genius of Edward Forbes.

6.
LOUIS AGASSIZ: "AN INTERCOURSE WITH THE HIGHEST MIND"

In July 1873, Louis Agassiz opened the first seaside laboratory in America, the Anderson Summer School of Natural History. With the greatest earnestness he told his students: "Our object is to study nature, and I hope I may lead you in this enterprise so that you may learn to read for yourselves. We should make nature our textbook; whenever we read books we are removed from the things we could be better acquainted with; instead of the things themselves we appropriate the interpretation of someone else; and, however correctly we may have done this, we invariably return to the study of the things themselves whenever we wish to make real progress; and I hope to live long enough to make textbooks useless and hateful, without even implying a reflection upon the services textbooks may have rendered in past times."

Thus spoke Louis Agassiz in his sixty-sixth year and at the height of world acclaim as a naturalist and teacher. The laboratory, on Penikese Island opposite New Bedford in Buzzards Bay, Massachusetts, was designed to help school and college teachers improve their methods of presenting natural history to their students. To attend this first summer session, Agassiz had chosen twenty-eight men and sixteen women from a long list of outstanding candidates. Agassiz's strength was no longer the marvel that it once had been, but he gave completely of himself to his lectures and demonstrations—as he had throughout his lifetime to projects so ambitious that most men would have quailed.

His character was a baffling mixture of monumental egoism and genuine humility. He regarded himself as so closely attuned to nature, and to the deity as he conceived it, that it was unthinkable to him that he might be wrong. In contrast, he would tell his students, "Have the courage to say: 'I do not know,' " and he would often follow that advice himself. When he was once asked what he regarded as his greatest work, he simply said, "I have taught men to observe."

Teaching men and women to observe was the central core of his work at the Penikese Island laboratory. But he maintained a running commentary for his students on the philosophy of science and the discipline of work. Many of them became celebrities in their own right. David Starr Jordan of Stanford University fame was one, and he took extensive notes of Agassiz's remarks:

"Lay aside all conceit," Agassiz would say. "Learn to read the book of nature for yourself. Those who have succeeded best have followed for years some slim thread which has once in a while broadened out and disclosed some treasure worth a life-long search." And again: "The study of nature is an intercourse with the highest mind. You should never trifle with nature. At the lowest her works are the works of the highest powers—the highest something, in whatever way we may look at it."

The subjects that Agassiz took up at the seaside were by no means limited to living plants and animals, for "the Professor" took the whole of nature as his province. Through many years of intense study—often extending to fifteen hours a day—he had made himself a world authority on geology, especially with respect to the actions of glaciers; on paleontology, especially with respect to fossil fishes; and on ichthyology, especially with respect to the fresh-water fishes of European lakes and streams. And, as always, Agassiz made the best use of the materials right at hand. With the Atlantic at the laboratory's front door, he saw to it that daily dredgings were carried out, and that each of his students studied some marine creature—usually an invertebrate—in the most minute detail. "It is much more important," he would say, "for a naturalist to understand the structure of a few animals than to command the whole field of scientific nomenclature."

Jean Louis Rodolphe Agassiz, always known as Louis from his childhood, was born of Swiss middle-class parents of modest income in 1807 in Môtier, a village of five hundred on the western shore of the Lake of Morat. Louis had little encouragement, either from his parents or from the local culture, to become a naturalist. The culture was one of artisans, tradesmen, and merchants, and the admired virtues were those of industry and thrift. The highest goal to which a young man might aspire was an adequate (preferably a generous) income, and the way lay either through business or through one of the professions. Agassiz's parents leaned toward business, for there were some very successful business people in the family; however, Louis's father looked on medicine as an acceptable alternative.

What force, then, attracted Agassiz so irresistibly to a lifelong study of nature?

A man of such complexity presents equally complex problems, but we can surely see two principal influences. One of these must lie in the beauty and richness of the region in which Louis was born and raised. The other lies in the inherent inner fiber of the man himself.

Môtier, in the Vully, is surrounded on three sides by water. To the west is the Lake of Neuchâtel, to the north is the river La Broye, and to the east is the Lake of Morat. North of all of these lies the *Seeland* of Berne, a wide sheet of marshy water. Behind these to the west, the glaciated Jura Mountains rise to five thousand feet and more, and open to Louis's view to the east, the Jungfrau capped the Bernese Alps. Valleys and meadows teemed with animals, lakes and rivers with many species of fishes, mountains with tumbled granite boulders and glaciated grooves that opened up fascinating clues to the history of creation.

To young Louis this was a treasure house for exploring, collecting, and enjoyment, and he made the most of it. As he was to note, "At that age, namely about fifteen, I spent most of the time I could spare from classical and mathematical studies in hunting the neighboring woods and meadows for birds, insects, and fresh-water shells. My room became a little menagerie, while the stone basin under the fountain in our yard was my reservoir for all the fishes I could catch. Indeed, collecting, fishing, and raising caterpillars, from which I reared fresh, beautiful butterflies, were then my chief pastimes."

And just pastimes they were, for Agassiz did not neglect his studies; indeed the family records included a time-worn sheet of foolscap on which he had set down—at the age of fourteen—a ten-year plan for his future intellectual development.

"I wish," he wrote, "to advance in the sciences, and for that I need d'Anville, Ritter, an Italian dictionary, a Strabo in Greek, Mannert and Thiersch; and also the works of Malte-Brun and Seyfert. I have resolved, as far as I am allowed to do so, to become a man of letters, and at present I can go no further; 1st, in ancient geography; for I already know all my notebooks, and I have only such books as Mr. Rickly [his teacher at the preparatory Collège de Bienne] can lend me; I must have d'Anville or Mannert; 2d, in modern geography, also, I have only such books as Mr. Rickly can lend me, and the Osterwald geography, which does not accord with the new divisions; I must have Ritter or Malte-Brun; 3d, for Greek I need a new grammar, and I shall choose Thiersch; 4th, I have no Italian dictionary, except one lent me by Mr. Moltz; I must have one; 5th, for Latin I need a larger grammar than the one I have, and I should like Seyfert; 6th, Mr. Rickly tells me that as I have a taste for geography he will give me a lesson in Greek (gratis), in which we would translate Strabo, provided I can find one. For all this I ought to have about twelve louis.[1] I should like to stay at Bienne until the month of July, and afterward serve my apprenticeship in commerce at Neuchâtel for a year and a half. Then I should like to pass four years at a university in Germany, and finally finish my studies in Paris, where I would stay about five years. Then, at the age of twenty-five, I could begin to write."

These teen-age notes clearly foreshadowed the man he was to become. He did begin to write, but well before he was twenty-five, and in his lifetime he published more than 425 articles and books. He never served his "apprenticeship in commerce," but he did study in German universities and in Paris, though not exactly in the order that he planned. Throughout his life he was careless of financial matters, always confident that the money to carry out his projects would somehow be forthcoming. Thus, hardly knowing from one month to the next where his food would come from (he paid scant attention to his clothes), he spent two years at Zurich, a year at the University of Heidelberg, and three years at the University of Munich. He emerged, thanks to a rugged constitution, as

---

[1] He did not lay hands on the twelve louis then; and even years later, he and his brother Auguste were copying out books in longhand that they could not afford to buy.

a doctor of philosophy in natural history, a doctor of medicine, and the author of a fairly important volume on the fresh-water fishes of Central Europe.

It was then high time, his parents thought, for him to plan to support himself. The sensible course, they argued, would be for him to return to Switzerland and establish a medical practice, on which he could rely. Or, failing this, to obtain a professorship in natural history that would at least bring him a steady income. But Agassiz's passion for natural history was unabated; besides, the suffering of sick people distressed him, and he avoided it as much as he could. Still, he continued to feel a strong obligation to "be a good son" and accede to his parents' wishes.

An epidemic of cholera that struck Paris in the early 1830's presented an unexpected possibility. Agassiz told his parents that every physician ought to learn as much as he could about cholera, and that Paris now offered an ideal place to study it. He probably sincerely intended to do so, but the magnet that really drew him to Paris was the great Muséum National d'Histoire Naturelle. At that time the Muséum was the outstanding natural-history center of Europe; it had a special building devoted to anatomy and paleontology; and Baron Georges Cuvier, the grand old man of comparative anatomy and paleontology, was a revered professor there.

With his usual self-confidence, backed by his prodigious capacity for work, he soon became Cuvier's protégé in the study of fossil fishes. Cuvier had originally intended to write the definitive book on the subject, but by 1831 he was already sixty-two, and he knew that the project would take many years. When he saw the skill and energy that Agassiz was devoting to his own study of fossil fishes, he turned all of his notes and specimens over to the younger man and wished him success in the venture.

A lavish feast for the mind was thus spread before Agassiz each day, but he could seldom afford good meals. His father could not understand what drove him to such extremes, and wrote a stream of letters urging him to leave Paris and find some kind of reliable employment. Agassiz tried to explain the importance of the work that he was doing. "The aim," he wrote in a carefully worded letter, "of our researches upon fossil animals is to ascertain what beings have lived at each one of these [geological] epochs of creation, and to trace their characters and relations with those now living; in one word to make them live again in our thoughts. It is especially the fishes that I try to restore for the eyes of the curious, by showing them which ones have lived in each epoch, what were their forms, and, if possible, by drawing some conclusions as to their probable modes of life. You will better understand the difficulty of my work when I tell you that in many species I have only a single tooth, a scale, a spine, as my guide to the reconstruction of all these characters, although sometimes we are fortunate enough to find species with the fins and skeletons complete."

Agassiz's fossil fishes "lived again" not only in his thoughts but also in his dreams. At one point he had been puzzled for weeks about the complete form of a fossil fish of which he had only fragments exposed in stone. On two successive nights he dreamed that he saw the complete fish, but when he awoke, he could not remember the details. On the third night, he laid pencil and paper by his bedside, and when he dreamed of the fish again, he sleepily sketched it. The next day he returned to the museum, and working with the utmost care, he disengaged his fish from its stony substrate. Unlike the exposed fragmentary side, the lower side of the fossil was complete—and it was an exact match for the sketch of his dreams!

Meanwhile, Agassiz's mother grew more and more concerned about his financial straits, and in March 1832 she wrote to him at length:

"It seems to me, my dear child, that you are painfully straitened in means. I understand it by personal experience, and in your case I have foreseen it; it is the cloud which has always darkened your prospect to me. I want to talk to you, my dear Louis, of your future, which has always made me anxious. . . . With much knowledge, acquired by assiduous industry, you are still at twenty-five years of age living on brilliant hopes, in relation, it is true, with great people, and known as having distinguished talent. Now, all this would seem to me delightful if you had an income of fifty thousand francs [Louis had just turned down an offer of *one* thousand francs to edit a zoological bulletin, because that would cost him two or three hours' time a day]; but, in your position, you must absolutely have an occupation which will enable you to live, and free you from the insupportable weight of dependence on others. . . . You must therefore leave Paris for Geneva, Lausanne, or Neuchâtel, or any city where you can support yourself by teaching. . . . It is a great evil to be spending more than one earns. . . ."

Two months later Cuvier died of the cholera that Agassiz had ostensibly gone to study. Nothing, then, held him to his near-destitute existence in Paris, and he began to think of the life he might lead as a professor, with a dependable income, and the time to study and write about fossil fishes.

Again luck came his way at a critical moment. The aristocracy of Neuchâtel and the Prussian monarchy had jointly decided that a new *collège* and museum of natural history should be established in the town. Money was forthcoming from both sources to establish a professorship in natural history, which would also entail the directorship of the museum. After some typically adroit maneuvering, Agassiz was awarded the appointment, and he settled down to a future—as he thought—of research and writing.

And he did indeed do extensive work in both. In the ten years that he spent at Neuchâtel, he completed his major scientific work, the *Recherches sur les poissons fossiles,* which was published in five lavishly illustrated parts during the course of the decade, and completed in 1843. In it he described and analyzed more than 1700 species of ancient fishes, and, as Edward Lurie said in *Louis Agassiz: A Life in Science,* it became "the primary inspiration for a new area of inquiry into the story of organic creation."

That was really the end of his original scientific work, for he was immediately drawn into a far broader contribution than any purely scientific work that he could have done. He accomplished, in short, a revolution in education.

The accepted method of teaching at that time was to require students to commit textbooks to memory and to recite them on demand. At Neuchâtel, Agassiz broke from this method and began experimenting with the novel scheme of asking his students to study directly, without the aid of any written guides, plants and animals and glacial rocks as they lay immediately at hand. His students progressed rapidly, buoyed by the adventure of discovering new things for themselves, and Agassiz was sure that he was heading in the right direction. He followed it throughout his life, and often in the next decades he would say to his students, "Study nature, not books."

After an illustrious decade at Neuchâtel, Agassiz was called to Boston. Again, characteristically, he went on a shoestring, but with supreme confidence that his fortunes would take care of themselves. The only firm commitment that he had to start with was a course of lectures to be delivered at the Lowell Institute on "The

Plan of the Creation, Especially in the Animal Kingdom."[2] Tickets to the Lowell lectures were drawn by lot, and were in great demand; men of letters sought them, as did men of science and business. But working men and women had equal chances at the drawings, and they thronged to hear Agassiz as enthusiastically as the "more fashionable." This was Agassiz's first exposure to the democratic principle in action, and he loved it. His joy in teaching burgeoned, his listeners were entranced, and Boston acquired a new legend.

As a result of the popularity of the Lowell series, Agassiz was offered a professorship in zoology and geology at Harvard, and at once he began assembling a vast collection of specimens from far-flung sources; upon that foundation he built the world-famous Museum of Comparative Zoology at Harvard. At the same time his winning personality and brilliant conversation attracted a wide circle of friends, and he was intimate with Asa Gray, Longfellow, Hawthorne, Holmes, Emerson, and Lowell. He continued to develop his revolutionary teaching method by locking students up with one bird, or one fish, or one insect, and requiring them to learn everything they could about it through direct observation, and to report to him in full detail.

Samuel H. Scudder, who was later to become a noted professor of entomology and a member of the Museum staff, gave an account of his own student experience that is so revealing that I quote it in full:

> It was more than fifteen years ago [from 1874] that I entered the laboratory of Professor Agassiz, and told him that I had enrolled my name as a student of natural history. He asked me a few questions about my object in coming, my antecedents generally, the mode in which I afterwards proposed to use the knowledge I might acquire, and, finally, whether I wished to study any special branch. To the latter I replied that, while I wished to be well grounded in all departments of zoology, I purposed to devote myself specially to insects.

---

[2] Although Agassiz was a contemporary of Charles Darwin, he never accepted the doctrine of evolution; in fact, he opposed it violently. The reasons for this are deeply entangled in his own concept of religion, his exposure to the natural philosophy of the German universities that he attended, and the reverence for unadorned facts that he had absorbed from Cuvier. The problem is too complex for our consideration here, but interested readers will find it brilliantly analyzed by Ernst Mayr in his essay "Agassiz, Darwin, and Evolution," *Harvard Library Bulletin*, Vol. XIII (Spring, 1959), pp. 165–94.

"Take this fish," said he, "and look at it; we call it a haemulon; by and by I will ask you what you have seen."

"When do you wish to begin?" he asked.

"Now," I replied.

This seemed to please him, and with an energetic "Very well!" he reached from a shelf a huge jar of specimens in yellow alcohol.

"Take this fish," said he, "and look at it; we call it a haemulon; by and by I will ask you what you have seen."

With that he left me, but in a moment returned with explicit instructions as to the care of the object entrusted to me.

"No man is fit to be a naturalist," said he, "who does not know how to take care of specimens."

I was to keep the fish before me in a tin tray, and occasionally moisten the surface with alcohol from the jar, always taking care to replace the stopper tightly. Those were not the days of ground-glass stoppers and elegantly shaped exhibition jars; all the old students will recall the huge neckless glass bottles with their leaky, wax-besmeared corks, half eaten by insects, and begrimed with cellar dust. Entomology was a cleaner science than ichthyology, but the example of the Professor, who had unhesitatingly plunged to the bottom of the jar to produce the fish, was infectious; and though this alcohol had a "very ancient and fishlike smell," I really dared not show any aversion within these sacred precincts, and treated the alcohol as though it were pure water. Still I was conscious of a passing feeling of disappointment, for gazing at a fish did not commend itself to an ardent entomologist. My friends at home, too, were annoyed, when they discovered that no amount of eau-de-Cologne would drown the perfume that haunted me like a shadow.

In ten minutes I had seen all that could be seen in that fish, and started in search of the Professor—who had, however, left the Museum; and when I returned, after lingering over some of the odd animals stored in the upper apartment, my specimen was dry all over. I dashed the fluid over the fish as if to resuscitate the beast from a fainting-fit, and looked with anxiety for a return of the normal sloppy appearance. This little excitement over, nothing was to be done but to return to a steadfast gaze at my mute companion. Half an hour passed—an hour—another hour; the fish began to look loathsome. I turned it over and around; looked it in the face—ghastly; from behind, beneath, above, sideways, at a three-quarters' view—just as ghastly. I was in despair; at an early hour I concluded that lunch was necessary; so, with infinite relief, the fish was carefully placed in the jar, and for an hour, I was free.

On my return, I learned that Professor Agassiz had been at the Museum, but had gone, and would not return for several hours. My fellow-students were too busy to be disturbed by continued conversation. Slowly I drew forth that hideous fish, and with a feeling of desperation again looked at it. I might not use a magnifying glass; instruments of all kinds were interdicted. My two hands, my two eyes, and the fish; it seemed a most limited field. I pushed my finger down its throat to feel how sharp the teeth were. I began to count the scales in the different rows, until I was convinced that that was nonsense. At last a happy thought struck me—I would draw the fish; and now with surprise I began to discover new features in the creature. Just then the Professor returned.

"This is right," said he; "a pencil is one of the best of eyes. I am glad to notice, too, that you keep your specimen wet, and your bottle corked."

With these encouraging words, he added:

"Well, what is it like?"

He listened attentively to my brief rehearsal of the structure of parts whose names were still unknown to me: the fringed gill-arches and movable operculum; the pores of the head, fleshy lips and lidless eyes; the lateral line, the spinous fins and forked tail; the compressed and arched body. When I had finished, he waited as if expecting more, and then, with an air of disappointment:

"You have not looked very carefully; why," he continued more earnestly, "you haven't even seen one of the most conspicuous features of the animal, which is as plainly before your eyes as the fish itself; look again, look again!" and left me to my misery.

I was piqued; I was mortified. Still more of that wretched fish! But now I set myself to my task with a will, and discovered one new thing after another, until I saw how just the Professor's criticism had been. The afternoon passed quickly; and then, toward its close, the Professor inquired:

"Do you see it yet?"

"No," I replied, "I am certain I do not, but I see how little I saw before."

"That is next best," said he earnestly, "but I won't hear you now; put away your fish and go home; perhaps you will be ready with a better answer in the morning. I will examine you before you look at the fish."

This was disconcerting. Not only must I think of my fish all night, studying, without the object before me, what this unknown but most visible feature might be; but also, without reviewing my new discoveries, I must give an exact account of them the next day. I had a bad memory; so I walked home by the Charles River in a distracted state, with my two perplexities.

The cordial greeting from the Professor the next morning was reassuring; here was a man who seemed to be quite as anxious as I that I should see for myself what he saw.

"Do you perhaps mean," I asked, "that the fish has symmetrical sides with paired organs?"

His thoroughly pleased "Of course! Of course!" repaid the wakeful hours of the previous night. After he had discoursed most happily and enthusiastically—as he always did—upon the importance of this point, I ventured to ask what I should do next.

"Oh, look at the fish!" he said, and left me again to my own devices. In a little more than an hour he returned, and heard my new catalogue.

"That is good, that is good!" he repeated; "but that is not

all; go on," and so for three long days he placed that fish before my eyes, forbidding me to look at anything else, or to use any artificial aid. "Look, look, look," was his repeated injunction.

This was the best entomological lesson I have had—a lesson whose influence has extended to the details of every subsequent study; the legacy the Professor has left me, as he has left it to many others, of inestimable value, which we could not buy, with which we cannot part.

A year afterward, some of us were amusing ourselves with chalking outlandish beasts on the Museum blackboard. We drew prancing starfishes; frogs in mortal combat; hydraheaded worms; stately crawfishes, standing on their tails, bearing aloft umbrellas; and grotesque fishes with gaping mouths and staring eyes. The Professor came in shortly after, and was as amused as any at our experiments. He looked at the fishes.

"Haemulons, every one of them," he said; "Mr. ——— drew them."

True; and to this day, if I attempt a fish, I can draw nothing but haemulons.

The fourth day, a second fish of the same group was placed beside the first, and I was bidden to point out the resemblances and the differences between the two; another and another followed, until the entire family lay before me, and a whole legion of jars covered the table and surrounding shelves; the odor had become a pleasant perfume; and even now, the sight of an old, six-inch worm-eaten cork brings fragrant memories.

The whole group of haemulons was thus brought in review; and, whether engaged upon the dissection of the internal organs, the preparation and examination of the bony framework, or the description of the various parts, Agassiz's training in the method of observing facts and their orderly arrangement was ever accompanied by the urgent exhortation not to be content with them.

"Facts are stupid things," he would say, "until brought into connection with some general law."

At the end of eight months, it was almost with reluctance that I left these friends and turned to insects; but what I had gained by this outside experience has been of greater value than years of later investigation in my favorite groups.

The establishment of the laboratory on Penikese Island still lay ahead. It was to be the last of Agassiz's ambitious projects for which he saw no means when he conceived it but for which he was sure he could find sponsors. In March 1873, the Massachusetts State Legislature paid its annual visit to the Harvard Museum of Comparative Zoology, and Agassiz seized the opportunity to make an impassioned plea for funds. The speech was printed in *The New York Times,* and John Anderson, a wealthy New York tobacco dealer, happened to read it. Within a week, he offered Agassiz the whole of Penikese Island, together with a barn, which would do for a lecture room and laboratory, a furnished house, and $50,000 for equipment. By July, Agassiz had opened for instruction the first seaside laboratory in America.

But his health was failing, and one morning in early December he complained of a "great weariness" and of dimness of sight, and he said that he felt "strangely asleep." He was confined to his bed for eight days, while his friends and family kept watch through an open doorway. At the last moment he sat up in his bed and exclaimed very distinctly, *"Le jeu est fini!"* Then he lapsed back, and by ten in the evening he was gone.

What pride and joy it would have given him if he could have seen the sequel to the Penikese Island summer school. Although it did not last very long, it was the inspiration and the direct predecessor of the world-famous Marine Biological Laboratory at Woods Hole. Founded in 1888, Woods Hole has attracted noted marine scientists from all over the world, and has been the birthplace of some of the most notable advances in marine science.

What pride and joy, too, it would have given him if he could have seen how widely his revolutionary teaching methods were adopted. Even outside science, they have been used with notable success. At Cornell, for example, Lane Cooper used them in graduate courses in English language and literature!

The psychologist William James summarized the Agassiz character and the Agassiz influence on a generation of Harvard students and faculty. Speaking at a reception of the American Society of Naturalists given by the President and Fellows of Harvard College on December 30, 1896, James said:

"The secret of such an extraordinarily effective influence lay in the equally extraordinary mixture of the animal and social gifts, the intellectual powers and the desires and passions of the man. From his boyhood he looked on the world as if it and he were made for each other, and on the vast diversity of living things as if he were there with authority to take mental possession of them all. His habit of collecting began in childhood, and during his long life he knew no bounds save those that separate the things of nature from those of human art. Already in his student years, in spite of the most stringent poverty, his whole scheme of existence was that of one predestined to greatness, who takes that fact for granted, and stands forth immediately as a scientific leader of men.

"His passion for knowing living things was combined with a rapidity of observation and a capacity to recognize them again and remember everything about them, which all his life it seemed an easy triumph and delight for him to exercise, and which never allowed him to waste a moment in doubts about the commensurability of his powers with his tasks. If ever a person lived by faith, he did."

James himself had been a student of Agassiz's, as had David Starr Jordan of Stanford and, of course, Samuel H. Scudder of the Harvard Museum. In addition to these, we find that nearly seventy Agassiz students made names for themselves in fields as varied as general zoology, psychology, mammalogy, ornithology, ichthyology, education, scientific authorship, entomology, agriculture, horticulture, archeology, geology, and the administration of scientific foundations.

All of these men and more were in the mind of James Russell Lowell when, from Italy, he wrote the following lines about Agassiz:

> *The beauty of his better self lives on*
> *In minds he touched with fire, in many an eye*
> *He trained to Truth's exact severity;*
> *He was a teacher: why be grieved for him*
> *Whose living word still stimulates the air?*
> *In endless file shall loving scholars come*
> *The glow of his transmitted touch to share.*

7.
MAYER: THE FORCES OF LIFE

The marine sciences offer some of the highest adventures that we may know. There is the anticipation as a dredge comes up from the depths of unexplored waters and we wait to see what treasures it may hold. There are the brilliant reds and yellows and iridescent blues of the creatures on a coral reef, to be seen with even the simplest diving goggles. There is the quiet satisfaction of a tide pool as we watch the amusing antics of the scurrying hermit crabs.

There are also high adventures of the human mind, as it searches for the general laws of nature that govern the lives of these innumerable creatures. Each of these living forms uses energy and performs work, and one of the most fundamental of all the laws of nature tells us what happens when this occurs. It is called the law of conservation of energy; behind it lies one of the most unusual stories in the entire history of science.

Sunlight powers the world. It strikes the earth as radiant energy, and as it strikes it is changed into a bewildering variety of other forms. It lifts the waters of the oceans high above the earth, providing a store of potential energy in the clouds. It appears in awesome forms in hurricanes and lightning bolts. It is also a form of chemical energy, stored silently by the green plants of the oceans—chemical energy to be transformed in many ways to serve the energy needs of man. Each of these many pathways—whether in living or in non-living systems—conforms to the law of conservation of energy.

The flashing of a firefly on a hot summer night, the flow of electricity from the generators of Hoover Dam, the surging power of a swordfish, all conform to this same universal law.

At the dawn of the nineteenth century the germ of the law lay just beneath the surface of scientific consciousness, for even then it was realized that some driving force must be behind the works of nature. What then was its source? Science did not know. The naturalists of the day called it *vis viva*, the force of life, but giving it a Latin name did not help them understand it any better.

Indeed, the period from 1800 to 1830 was one of fumbling, stumbling, and occasional outrageous error. But beginning in about 1830, the scientific community began to sense the glimmer of an idea, and by 1850, the greatest and most fundamental of all the laws of nature had been fully developed and generally accepted. This law states that in any system that neither gains nor loses energy to the outside world, the amount of energy that it contains is constant. This energy can be changed from one form to another in an unending sequence of events, but whatever form it takes, it is neither created nor destroyed.

More than a dozen scientists made significant contributions to the final concept. Outstanding among them are four: Herman von Helmholtz, Ludwig A. Colding, James Joule, and Julius Robert von Mayer, usually known simply as Robert Mayer. Mayer was the first to publish, though not the first to be recognized, for he was not at the time well known. But it was Mayer who made at once a great leap of the imagination and a far-reaching interpretation of energy conversions throughout the universe. And his concepts have needed little change since then.

Did the scientific community then rush to accept the new doctrine? It did not. Mayer even had difficulty in getting his work published, and when he did, no one paid much attention to it.

One of his difficulties was that he held no degree in natural history and no well-known professional chair. His formal training was in medicine, and he continued a rather obscure medical practice throughout most of his life. His attraction to science came about through a curious accident, and he never pursued experimental work in science as more than a sideline.

Robert Mayer was born at Heilbronn, Germany, in 1814, studied medicine at the University of Tübingen, and was admitted to medical practice at the age of twenty-four. However, he was not yet ready to settle down, and he signed on as surgeon on a ship bound for the South Seas. Halfway around the world, in Java, Mayer was called to treat an urgent case—a patient suffering from a raging fever. Although he had no knowledge of tropical medicine, and could not identify the cause of the fever, Mayer knew that the first thing to do was to break the fever. The procedure of choice at the time was to bleed the patient, and Mayer immediately lanced a vein and drained blood into a basin. His first concern was for the patient, and he watched him closely. Then, as he cut off the flow of blood and bound up the incision, he was startled by the color of the blood. In contrast to the typical blue of venous blood, this had a pronounced reddish tinge.

What could possibly cause that? How did this patient differ from all the others that he had attended? The most obvious difference was in climate. This man had lived in the tropics all his life. Was it possible that the blood of a Javanese might be different from that, say, of an Eskimo? If so, why?

Mayer was familiar with Lavoisier's work on animal metabolism, and he knew that the blood absorbs oxygen in the lungs, and that this changes its color from venous blue to arterial red. In the tropics, he reasoned, a man would easily retain his body heat, and would therefore need less oxygen than a man who lived in a cold climate. Could this be why the venous blood was redder?

At the end of the voyage, Mayer returned to Heilbronn and established himself as town physician, but he could not put the Java incident out of his mind. He considered the problem from many different angles, and he asked himself how else a man used energy besides maintaining his body temperature. The obvious answer was that he worked. With that, Mayer made his great imaginative leap. A man's body used the energy of food to keep its temperature up. It also used energy to perform work—a very different thing. Might there then be some fixed relationship between heat and work? If so, it ought to be possible to measure it, and that is just what Mayer determined to do.

Before we join him in his study, however, let us look more closely at the implications of the terms "energy" and "work." We know in a general way what these mean, but to appreciate Mayer's contribution fully, we need to pin them down as precisely as we can.

First, heat is simply one form of energy, and we must therefore seek an answer in the meaning of energy. This is a troublesome problem philosophically, for it can only be said that energy is a *capacity;* we cannot say what it really is. But, fortunately, we can say with confidence what it does. Energy is the capacity for doing work or, more specifically, it is the capacity for changing the motion of matter. The energy in a baseball player's bat has the capacity to change the motion of a high fast ball from a strike to a home run. In doing so, it performs work, which can be expressed as the product of the force exerted on the ball and the distance it travels.

But these are modern definitions. When Mayer conceived his experiment, the scientific community had no clear-cut idea of what any of these terms meant. Many scientists still regarded heat as a fluid, which they called "caloric," rather than the motion of molecules that we now know it to be. And many thought of energy as the old *vis viva.* In short, there were no textbooks and no authorities to which Mayer could turn. He was completely on his own. That, perhaps, is why he had a revolutionary idea.

After he had it, he confirmed it by skillful laboratory experiments, but the drama in this story is not of the laboratory, but of the mind. Let us therefore follow Mayer's thoughts as he worked out the plan of the experiment that he would afterward perform.

He decided that he would first take a certain weight of a gas—say, the weight of a gas that would fill a quart cylinder at room temperature and ordinary atmospheric pressure. He would then heat that gas in two separate experiments under two entirely different sets of conditions. First he would heat the gas in a closed vessel so that it could not expand, and measure exactly the number of calories that it would take to raise the temperature of the gas by, say, ten degrees. This would be a function of the specific heat (the heat-holding capacity) of the gas.

For the second experiment, he would construct a cylinder sealed at the top by a gas-tight piston that could slide up and down as the gas expanded or contracted. Of course, he would know the exact weight of the piston.

Now, when a gas is heated, it tends to expand, and it will do so if it can force a movable piston upward. Mayer planned to heat the same weight of the same gas through the same ten-degree temperature rise, but simultaneously to allow the gas to drive the piston up. Exactly the same number of calories would be needed to raise the temperature of the gas, but now more would be needed to lift the piston. In short, the difference in the heat inputs in the two experiments would be equivalent to the work that the expanding gas performed on the piston. This difference would be a simple, direct measure of the mechanical equivalent of heat.

In this actual laboratory experiment Mayer did not take into account the fact that few gases are perfect, and so his value for the mechanical equivalent differed from the precise one that we know today. But even so, Mayer's value was close, and the inaccuracy does not detract in the least from the brilliance of the concept.

Viewed in another way, Mayer's experiment also clearly showed that heat alone can both raise the temperature of a body and perform work. This strongly supported his speculation that heat from slow internal combustion in our bodies is used partly to maintain our body temperature and partly to do our work.

Three years later, in 1845, Mayer carried out more refined experiments that yielded a better value for the mechanical equivalent of heat, and he greatly expanded his ideas about the variety of energy transformations that take place in the world.

Throughout all his work he held fast to his basic conclusion that energy can never be created or destroyed. An extension of this was that the long-sought perpetual motion machine was an impossibility; for such a machine would have to create energy in order to run. Curiously, this harked back to Mayer's boyhood, when he had made a hobby of model waterwheels; from a study of these, he had convinced himself even then that perpetual motion was impossible.

You will recall that the combined work of Priestley, Ingenhousz, de Saussure, and Sachs showed that green plants transform the radiant energy of sunlight into the chemical energy of carbohydrates. Now Mayer took a long stride forward in proposing that the sun was the sole source of all the energy on earth, whether it powered living or non-living systems, and he began to examine all the systems that he could think of, to see if he could find any exceptions.

He considered a gamut of biological processes, the surges of the tides, and the incandescence of meteors as they enter the earth's atmosphere and release their kinetic energy in the form of light and heat. And he argued that the energy for all these processes had its origin in the sun.

When, at last, he achieved publication of his work, he had some hope that the soundness of his arguments would be apparent to all. He could not have been more wrong. He failed to realize that a scientific discovery must fit the temper of the times, and the time was not yet ripe for the unknown Robert Mayer. Even though the illustrious Julius Liebig finally published the work, the paper languished.

How could this happen? Perhaps the most important reason is that truly revolutionary ideas have a hard time finding receptive ground. A scientific community at any given moment is wrapped up in its own pursuits—pursuits of what Thomas Kuhn has called "normal science" in his penetrating volume, *The Structure of Scientific Revolutions*. Revolutionary ideas disturb the orderly course of "normal science"; they rock the boat. No matter how sound they may eventually prove to be, new ideas must pass the test of time before an established scientific community will receive them. Unaware of this simple fact of life, Mayer chafed and fretted at his lack of recognition. He grew more and more quarrelsome, and the more hotly he argued, the more often he lost.

By 1849 he was so violently disturbed that he tried to commit suicide by jumping from a second-story window. That was poor judgment, for he failed there too. Within another two years he had to be shut up in a mental institution—a sorry place indeed in the middle of the nineteenth century. Although Mayer was eventually released, some thought that the harsh treatment that he received had permanently affected his mind.

But strangely, his closing years brought him the recognition that he had so long yearned for. Honors began to come his way, and in 1871 the Royal Society awarded him the Copley Medal. He spent his last days in his native town of Heilbronn, in the valley of the Neckar, surrounded by peaceful vineyards, where he found his own peace tending his own vineyard. Thirty years after his attempt on his own life, he died quietly on his native soil.

He lived to read the tribute paid him by Sir John Tyndall, Rumford Medalist and professor of natural history at the Royal Institution. Speaking before the Royal Society, Sir John had said: "Without external stimulus, and pursuing his profession as

On this early California mill wheel, the potential energy of water stored in a lake above was converted to the kinetic energy that ground wheat into flour.

This deep-sea angular fish, Himantolophius groenlandicus, *is about the size of a football. It attracts its prey through luminous organs in which chemical (food) energy is converted to light.*

town physician at Heilbronn [Robert Mayer] was the first to raise the conception of the interaction of heat and other natural forces to clearness in his own mind. And yet he is scarcely heard of, and even to scientific men his merits are but partially known. Led by his own beautiful researches, and quite independent of Mayer, Mr. Joule published in 1843 his first paper on the 'Mechanical Value of Heat'; but in 1842 Mayer had actually calculated the mechanical equivalent of heat from data which only a man of the rarest penetration could have turned to account. In 1845 he published his 'Memoir on Organic Motion,' and applied the mechanical theory of heat in the most fearless and precise manner to vital processes. He also embraced the other natural agents in his chain of conservation. In 1853 Mr. Waterson proposed, independently, the mechanical theory of the sun's heat, and in 1854 Professor William Thomson applied his admirable mathematical powers to the development of the theory; but six years previously the subject had been handled in a masterly manner by Mayer. . . . When we consider the circumstances of Mayer's life, and the period at which he wrote, we cannot fail to be struck with astonishment at what he had accomplished. Here was a man of genius working in silence, animated solely by a love of his subject, and arriving at the most important results in advance of those whose lives were entirely devoted to Natural Science."

Mayer's "Memoir on Organic Motion" indeed applied in a "fearless" manner to a wide variety of energy conversions in living systems, and in one way or another he anticipated many later discoveries. Let us briefly note some of these later ones now, for while it will be ahead of our story, mention of them here will put the whole concept of energy conversions into perspective. We now know a great deal about the following kinds of transformation of energy:

- The change of chemical energy to electrical energy in nerve cells and brain cells.
- The change of sound to electrical energy in the inner ear.
- The change of light energy to chemical energy in the chloroplasts of green plants.
- The change of light to electrical energy in the retina of the eye.
- The change of chemical energy to osmotic energy in the cells of the kidney.

When Kovachi, a submarine volcano in the Solomon Islands, erupted in October 1969, vast stores of potential energy within the earth were converted into the kinetic energy of shock waves, and spuming steam and spray.

- The change of chemical energy to mechanical energy in muscle cells.
- The change of chemical energy to light in the luminescent organs of many different marine organisms.
- The change of chemical energy to electrical energy in the sense organs of taste and smell.

These are but a few of the transformations of energy that are constantly occurring throughout the living world. If we include parallel transformations in non-living systems, we can regard the whole earth, and especially its oceans, as a gigantic heat engine, operating a complex of an infinite number and variety of sub-processes, every one of which obeys the greatest and most fundamental law of nature. A law, moreover, that was discovered by an obscure German town doctor who asked himself why the color of one sample of venous blood looked unusually red.

8.
THOMSON: THE DEPTHS OF THE SEA

Four days before Christmas in the year 1872, a flush-decked British corvette, powered by twelve hundred horses of auxiliary steam, left Portsmouth on a voyage that was to revolutionize man's knowledge of the sea. Stripped down to two of her normal eighteen heavy guns, so that her team of naturalists might have ample deck space for their work, H.M.S. *Challenger* laid her course for the Canary Islands, on the first leg of a voyage that was to explore 69,000 miles of the Atlantic, the Pacific, and the southern ice of the Antarctic. "Never," Sir William Herdman later said, "did an expedition cost so little and produce such momentous results for human knowledge."

That sailing date was the culmination of twenty years of dedicated effort on the part of a man who never completed his formal university education—Charles Wyville Thomson. Of course, Thomson was assisted in many ways by other capable and dedicated men, but he was always a step ahead along the historic chain of events that led up to that great day.

As chief of the scientific party aboard the *Challenger,* Wyville Thomson led studies of the surface currents of the ocean, studies of plankton and abyssal animals, of deep-sea deposits on the ocean floor, and of the life history of coral reefs. He also contributed much to the improvement of methods and equipment for exploring the depths of the open ocean.

The results of the *Challenger* expedition were eventually published in fifty large volumes, and marine biologists who specialize in particular orders of plants or animals must consult these books to this day.

Charles Wyville (later Sir Wyville) Thomson was born in March 1830, at Linlithgow, Scotland, near the shores of the Firth of Forth, where for generations his family had had intimate associations with Edinburgh and its neighborhood. Young Charles attended Merchiston Castle School, and while his favorite studies then were the Latin poets, it is worth noting that no science was taught there then.

When he was sixteen he began the study of medicine, but it was the natural sciences that absorbed his interest. These he studied intensively, but he never bothered to take a formal degree.

Today the lack of that academic union card would bar a man from any university teaching post, but a century ago that was not the case. Wyville Thomson made impressive progress in his individual studies in zoology, botany, and geology, and at the age of twenty-one he was appointed lecturer in botany at the University of Aberdeen. Two years later he went to Ireland as professor of natural history at Queen's College, Cork, and in 1854 he was appointed professor of geology at Belfast. In 1860 he was appointed professor of zoology and botany, also at Belfast.

In that post he spent a most rewarding decade. His lecture room was often jammed with admiring students, and one of them gave his impressions of that period to Sir William Herdman (himself at one time a student of Thomson's), who wrote the following vignette:

"Thomson had a bright, handsome face and a light, springy step; he was a delightful and instructive lecturer, who had on his table a profusion of specimens of which he made constant use, but spoke without notes. His Saturday excursions must have been delightful. We have a picture of him striding along, vasculum [specimen case] on back, at the head of his students, pointing out specimens and objects of interest as they were encountered. His hospitality to his students has left pleasant memories of the music

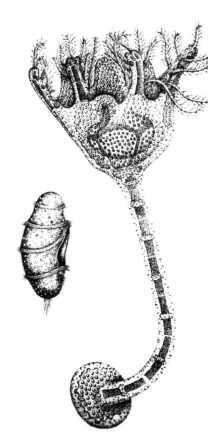

Strange pre-adult forms of a familiar starfish. The pseudo-embryo *(left) and the* pentacrinoid larva *(right) of Antedon rosaceus, both much magnified. Sir Wyville Thomson was struck by "the wonderful beauty of its form and movements . . . rendering it an object of ever-recurring admiration and interest."*

and games at their social evenings. Amongst other activities at Belfast, he took a prominent position at the Natural History and Philosophy Society, the Belfast Naturalists' Field Club, and also the Literary Society, at all of which he read papers. We hear that he gloried in his beautiful garden and was a valued judge at the local flower shows."

Side by side with his teaching duties at Belfast, Wyville Thomson carried on a series of field studies of British coelenterates (the animal group that includes polyps, jellyfishes, and most corals), and his interest in fossil forms sparked a three-year laboratory study of the crinoids—cup-shaped animals with branched radiating arms. In his studies of paleontology, Thomson had become interested in the fossil crinoids (such as sea lilies and feather stars) as possibly throwing some light on how living forms developed through geological time, for in the geological past crinoids were a dominant form; some five thousand fossil species have been described. Today only about six hundred living species are known, and the only representative of these in the British seas is *Antedon rosaceus,* the rosy feather star. What fascinated Thomson especially about this beautiful animal was that it emerges as a free-swimming larva that lives for a few days off the yolk of the egg, and then sits down, glues itself by an adhesive pit to some handy rock, and metamorphoses into a stalked sessile form known as the "pentacrinoid." "Just how the tail end of a delicate larva turns into a heavy five-rayed echinoderm [meaning having a skin like a hedgehog] is one of the most extraordinary metamorphoses in the animal kingdom."

In his Belfast laboratory Thomson set up a battery of aquarium tanks supplied with circulating sea water, and in them he carried out a painstaking study of *Antedon*'s early development from the fertilization of the egg (it has two sexes) through the transformation into the stationary "pentacrinoid" form. He presented the results of this study in a paper that he read before the Royal Society on February 5, 1863, and published later in thirty-one close-set pages (illustrated with exquisitely drawn figures) in the *Philosophical Transactions of the Royal Society* for 1865. With his friend William B. Carpenter of the University of London, he did some further work on the later development of the "pentacrinoid" form of *Antedon,* but he did not carry the paleontological implications much further. Wider vistas lay before him.

Thomson was an intimate and a great admirer of Edward Forbes, and had frequently dredged with him in the Firth of Forth. The two had often talked about Forbes' suggestion that life could not exist below three hundred fathoms, and Thomson accepted the proposition as self-evident. So indeed did most biologists of the day. We can easily see why. Below three hundred fathoms, they reasoned, no light could penetrate; the pressure of tons of water above would be unbearable; and the water itself would be far denser than at the surface.

Now, the assumption of Stygian darkness in the ocean deeps was right; we may be surprised at how shallow the penetration of sunlight really is. If you hold up a glass of water to the light, it seems crystal clear, but in the sea light rays are quite rapidly absorbed. Sunlight loses half its intensity at a depth of sixteen feet, and at seventy-five feet some 97 per cent or more of it has disappeared. Surely green plants that need sunlight to exist would not be found below this level. What animal could live in a watery waste barren of plant life?

The belief that enormous pressures prevail at great ocean depths was also right. No test was needed to prove that; even a young student of physics could quickly calculate that pressures on the ocean floor may be more than a ton per square inch. Wyville Thomson himself imagined an amusing picture. "At 2000 fathoms a man would bear upon his body a weight equal to twenty locomotive engines, each with a long goods train loaded with pig iron."

Under such enormous pressures, men further assumed, the water itself would be much compressed and would become so dense that heavy objects could not sink in it. All the loose things in the ocean would ride at different levels, depending on their own densities. At the shallowest would be porous rocks, for example, and below them would float the skeletons of men. Further down would be layers of anchors and cannon and shot, and deeper still would float broad gold pieces spilled from galleons lost on the Spanish Main. And in the deepest reaches of the sea would lie a layer of clear, pure water denser than molten gold.

We can see a striking parallel here with Aristotle's doctrine that a heavy body will fall faster than a light one "because it ought to." Of course, when Galileo actually tried it, he demonstrated that Aristotle was wrong. And when physicists in the laboratory showed that water is hardly compressible at all, Aristotle's latter-day counterparts allowed the skeletons, and cannon balls, and pieces of eight to sink to the bottom of the sea.

Then, from various parts of the world—from America and the Mediterranean and Norway—there came unsettling news. Living forms were being brought up from ocean depths very much greater than Forbes' three hundred fathoms. Even as early as 1818, Sir John Ross had found a remarkable medusa with branching arms entangled in a sounding line from a depth of eight hundred fathoms in Baffin Bay. It was a beautiful animal, some three inches across, with five double arms arising from the edge of a central disk. But the report did not attract much attention; men assumed that the medusa had just wandered there by chance.

In 1854, J. M. Brook, a young midshipman in the United States Navy, hit upon an ingenious device for taking samples of bottom ooze from the end of a sounding line. He fastened a quill to the lead and brought up tiny samples of oozy mud. How simple! And how exciting, for Brook's quills recovered microscopic shells of animals from all over the Atlantic, from depths of more than a thousand fathoms. Most of these were of one-celled forams of the genus *Globigerina*. Looking at a living specimen through a fifty-power microscope, we would see three roughly spherical balls of a deep vibrant red, closely bunched together, with a profusion of

A young Caput medusae *such as Sir John Ross found in 1818 at one thousand fathoms in Baffin Bay.*

fine-spun filaments radiating from each rosy ball like thistledown of dusty blue.

Again, in the summer of 1860 the Mediterranean cable between Sardinia and the African coast gave way at a depth of twelve hundred fathoms. Fleeming Jenkin, a civil engineer, was retained by the Mediterranean Telegraph Company to repair it. In the course of the operation Jenkin had to fish up some forty miles of cable that had been submerged at depths ranging from seventy to twelve hundred fathoms. Clinging to this entire length of cable were marine animals of many kinds, and at twelve hundred fathoms Jenkin found a true coral about an inch across, looking very much like a bent ice-cream cone filled with small mussel shells set on edge. Jenkin sent samples that he had scraped from the cable to Professor Allman, who had succeeded Edward Forbes at Edinburgh, and Allman identified fifteen animal species.

In the face of this mounting evidence, Wyville Thomson grew more and more skeptical of Edward Forbes' idea that life could not exist at three hundred fathoms. Then he learned that Michael Sars, a leading marine biologist in Norway, had found a number of new species of animals that had been dredged up near the Lofoten Islands from depths greater than four hundred and fifty fathoms. One of these was a strange new stalked crinoid, which was a close parallel to *Antedon,* with which Thomson was so familiar. Soon he was in Norway, exchanging ideas with Sars, and his vigorous imagination began to shape a plan for a new kind of deep-sea exploration—far more ambitious than anything that had ever been attempted.

Brimming with enthusiasm, he returned to Belfast and got in touch with his old friend William B. Carpenter in London. Soon Carpenter visited Belfast, and the two shared visions of the great scientific advances that might flow from a more systematic investigation of the ocean depths. As vice-president of the Royal Society, Carpenter was well placed to secure prestigious support for the plan (a circumstance that can hardly have escaped Thomson's notice), and when he returned to London, he sent off the following letter to the Royal Society:

> University of London,
> Burlington House, W.
> June 18, 1868.

Dear General Sabine,— During a recent visit to Belfast, I had the opportunity of examining some of the specimens (transmitted by Prof. Sars of Christiana to Prof. Wyville Thomson) which have been obtained by M. Sars, jun., Inspector of Fisheries to the Swedish Government, by *deep-sea* dredgings off the coast of Norway. These specimens ... are of singular interest alike to the zoologist and the paleontologist; and the discovery of them can scarcely fail to excite, both among naturalists and among geologists, a very strong desire that the zoology of the *deep sea,* especially in the Northern Atlantic region, should be more thoroughly and systematically explored than it has ever yet been. From what I know of your own early labours in this field, I cannot entertain a doubt of your full concurrence in this desire.

Such an exploration cannot be undertaken by private individuals, even when aided by grants from Scientific Societies. For dredging at great depths, a vessel of considerable size is requisite, with a trained crew, such as is only to be found in the government service. It was by the aid of such an equipment, furnished by the Swedish Government, that the researches of M. Sars were carried on.

Now, as there are understood to be at the present time an unusual number of gunboats and other cruisers on our northern and western coasts, which will probably remain at their stations until the end of the season, it has occurred to Prof. Wyville Thomson and myself, that the Admiralty, if moved thereto by the Council of the Royal Society, might be induced to place one of these vessels at the disposal of ourselves and of any other naturalists who might be willing to accompany us, for the purpose of carrying on a systematic course of deep-sea dredging for a month or six weeks of the present summer, commencing early in August.

Though we desire that this inquiry should be extended both in geographical range and in depth ... we think it preferable to limit ourselves on the present occasion to a request which will not, we believe, involve the extra expense of sending out a coaling vessel. We should propose to make Kirkwall or Lerwich our point of departure, to explore the sea-bottom between the Shetland and the Faeroe Islands, dredging around the shores and in the fiords of the latter (which have not yet, we believe, been scientifically examined), and then to proceed as far north-west into the deep water between the Faeroe Islands and Iceland as may be found practicable.

It would be desirable that the vessel provided for such

service should be one capable of making way under canvas as well as by steam-power; but as our operations must necessarily be slow, *speed* would not be required. Considerable labour would be spared to the crew if the vessel be provided with a 'donkey-engine' that could be used for pulling up the dredge.

If the Council of the Royal Society should deem it expedient to prefer this request to the Admiralty, I trust that they may further be willing to place at the disposal of Prof. Wyville Thomson and myself . . . a sum of £100 for the expenses we must incur in providing an ample supply of spirit and of jars for the preservation of specimens, with other scientific appliances. We would undertake that the choicest of such specimens should be deposited in the British Museum.

I shall be obliged by your bringing this subject before the Council of the Royal Society, and remain,

Dear General Sabine, Yours faithfully,
William B. Carpenter.

The Council of the Royal Society sent this request along with a strong second; the Admiralty agreed, and on August 8, 1868, H.M.S. *Lightning* left Oban for the Faeroes. Of this first of all deep-sea research ships Thomson wrote:

"The surveying ship 'Lightning' was assigned for the service —a cranky little vessel enough, one which had the somewhat doubtful title to respect of being perhaps the very oldest paddle-steamer in Her Majesty's Navy. We had not good times in the 'Lightning.' She kept out the water imperfectly, and as we had deplorable weather during nearly the whole of the six weeks we were afloat, we were in considerable discomfort. The vessel, in fact, was scarcely seaworthy, the iron hook and screw jack fastenings of the rigging were worn with age, and many of them were carried away, and on two occasions the ship ran some risk. Still the voyage was on the whole almost pleasant. Staff-Commander May had lately returned from Annesley Bay, where he had been harbour-master during the Abyssinian war; and his intelligence and vivacity, and the cordial good fellowship of his officers, who heartily seconded my colleague and myself in our work and sympathized with us in our keen interest in the curious results of the few trials at great depths which we had it in our power to make, made the experience a very novel one to us, certainly as tolerable as possible."

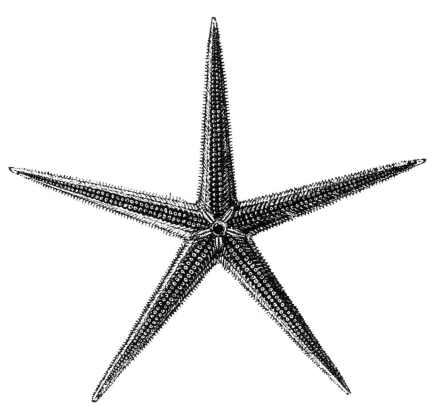

A new species of starfish, Archaster vexillifer, *discovered by Sir Wyville Thomson in the North Atlantic. Its color is pale rose touched with buff. It is about ten inches across.*

In spite of the *Lightning*'s crankiness and that summer's deplorable weather, Thomson and Carpenter dredged to depths of six hundred fathoms, and took some specimens of species unknown to science, and many species that had never before been recovered from such depths. In his timeless book, *The Depths of the Sea,* Thomson gives a detailed log of the operations and results at each dredging station. Summarizing the cruise, he wrote:

"The general results of the 'Lightning' expedition were upon the whole as satisfactory as we had ventured to anticipate. The vessel was certainly not well suited for the purpose, and the weather throughout the cruise was very severe. During the six weeks which elapsed between our departure from Oban and our return, only ten days were available for dredging in the open sea, and on four of these only were we in water over 500 fathoms deep. On our return Dr. Carpenter submitted to the Royal Society a preliminary report of the general results of the cruise, and these were regarded by the Council of the Society as sufficiently new and valuable to justify a strong representation to the Admiralty urging the importance of continuing the investigation which had already, under unfavourable circumstances, achieved a fair measure of success.

"It had been shown beyond question that animal life is varied and abundant, represented by all the invertebrate groups, at depths in the ocean down to 650 fathoms at least, notwithstanding the extraordinary conditions to which animals are there exposed."

New representations *were* made by the Council of the Royal Society, and the Admiralty agreed to provide another vessel, this time more suited to deep dredging. "The 'Porcupine,' though a small vessel, was well suited for the work; thoroughly seaworthy, very steady, and fitted up for surveying purposes." In the summer of 1869 she made successful hauls as deep as three miles down, and every haul from whatever depth teemed with life. One haul near Stornoway produced an unusually remarkable find.

"This haul was not very rich," Thomson wrote, "but it yielded one specimen of extraordinary beauty and interest. As the dredge was coming in we got a glimpse from time to time of a large scarlet urchin in the bag. We thought it was one of the highly-coloured forms of *Echinus flemingii* of unusual size, and as it was blowing fresh and there was some little difficulty in getting the dredge capsized, we gave little heed to what seemed to be an inevitable necessity—that it should be crushed to pieces. We were somewhat surprised, therefore, when it rolled out of the bag uninjured; and our surprise increased, and was certainly in my case mingled with a certain amount of nervousness, when it settled down quietly in the form of a round red cake, and began to pant— a line of conduct, to say the least of it, very unusual in its rigid undemonstrative order. Yet there it was with all the ordinary characters of a sea-urchin, its inter-ambulacral areas, and its ambulacral areas with their rows of tube feet, its spines, and five sharp blue teeth; and curious undulations were passing through its perfectly flexible leather-like test. I had to summon up some resolution before taking the weird little monster in my hand, and congratulating myself on the most interesting addition to my favourite family which had been made for many a day."

Though many of the dredgings were successful, some of them were not, and Thomson often found that while little came up inside the dredge, many echinoderms, corals, and sponges came up sticking to the outside of the bag.

"This," Thomson reported, "suggested many expedients, and finally Captain Calver sent down half-a-dozen 'swabs' used for washing the deck attached to the dredge. The result was marvelous. The tangled hemp brought up everything rough and moveable which came its way, and swept the bottom as it might have swept the deck. Captain Calver's invention initiated a new era in deep-sea dredging."

Thomson's enthusiasm for the hemp tangles was boundless. "I do not believe human dredger ever got such a haul. The special inhabitants of that particular region—vitreous sponges and echinoderms—had taken quite kindly to the tangles, warping themselves into them, and sticking through them and over them, till the mass was such that we could scarcely get it on board. Dozens of great *Holteniae*, like

'*Wrinkled heads and aged,*
With silver beard and hair,'

a dozen of the best of them breaking off just at that critical point where everything doubles its weight by being lifted out of the water, and sinking away back again to our inexpressible anguish; glossy wisps of *Hyalonema* spicules; a bushel of pretty little mushroom-like *Tisiphonia;* a fiery constellation of the scarlet *Astropecten tenuispinus;* while a whole tangle was ensanguined by the 'disjecta membra' of a splendid *Brisinga*."

So it was that two brief but arduous summer voyages within a small range of the British seas opened up a vast field for scientific investigation. The ocean deeps were not barren of life. They teemed with it! Around the Lews, the Orkneys, the Shetlands, and the Faeroes many new species were found, and the deeps of all the world's oceans cried out to be explored.

By the time he was barely forty, Thomson had written several classic papers for the Royal Society, his discoveries on the *Lightning* and the *Porcupine* were well known to naturalists throughout the world, and he was appointed Regius Professor of Natural History at the University of Edinburgh—the post that Edward Forbes had held fifteen years before. Sir William Herdman, who was a student of Thomson's and later his assistant, thought that if time and health had permitted, Thomson might have developed a truly great school of marine biology at Edinburgh. But he did not. His natural bent lay along more far-ranging lines.

The discoveries that he had made aboard the *Lightning* and the *Porcupine* only whetted his appetite for more, and by 1871 he had induced the Council of the Royal Society to seek Admiralty support for a far more ambitious voyage. Again, he had the wholehearted backing of Professor Carpenter of the University of London. Perhaps the key instrument that led to approval of the plan was the letter that the Royal Society sent to the Admiralty. It is a persuasive document and a model of its kind.

To the Secretary of the Admiralty.
The Royal Society, Burlington House,
December 8th, 1871.

Sir:— I am directed by the President and the Council of the Royal Society to request that you will represent to the Lords Commissioners of the Admiralty that the experience of the recent scientific investigations of the deep sea, carried on in European waters by the Admiralty at the instance of the Royal Society . . . has led them to the conviction that advantages of great importance to Science and Navigation would accrue from the extension of such investigations to the great ocean regions of the globe. The President and the Council therefore venture to submit to their Lordships' favourable consideration a proposal for fitting out an Expedition commensurate to the objects in view; which objects are briefly as follows:—

(1) The physical conditions of the deep sea throughout all the great ocean-basins.

(2) The chemical constitution of the water at various depths from the surface to the bottom.

(3) The physical and chemical characteristics of the deposits.

(4) The distribution of organic life throughout the areas explored.

For effectively carrying out these researches there would, in the opinion of the President and the Council, be required—

(1) A ship of sufficient size to afford accommodation and storage-room for sea-voyages of considerable length and for probable absence of four years.

(2) A staff of scientific men qualified to take charge of the several branches of the investigation.

(3) A supply of everything necessary for the collection of the objects of research, for the prosecution of the physical and chemical investigations, and for the study and preservation of the specimens of organic life.

The President and Council hope that in the event of their recommendation being adopted, it may be possible for the Expedition to leave England some time in the year 1872; and they would suggest that as its organization will require much time and labour, no time should be lost in the commencement of the preparations.

The President and the Council desire to take this opportunity of expressing their readiness to render every assistance in their power to such an undertaking; to advise upon (1) the route which might be followed by the Expedition, (2) the scientific equipment, (3) the composition of the scientific staff, (4) the instructions for that staff; as well as upon any matter connected with the Expedition upon which their Lordships might desire an opinion.

The President and Council have abstained from any allusion to geographical discovery or hydrographic investigations, for which the proposed Expedition would doubtless afford abundant opportunity, because their Lordships will doubtless be better judges of what may be conveniently undertaken in these respects, without departing materially from the primary objects of the voyage; and they would only add their hope that, in accordance with the precedents followed by this and other countries in somewhat similar circumstances, a full account of the voyage and its scientific results may be published under the auspices of the Government as soon after its return as convenient, the necessary expenses being defrayed by a grant from the Treasury.

The President and the Council desire, in conclusion, to express their willingness to assist in the preparation of such publication of the scientific results.

I remain, &c.

Again the Admiralty agreed to the Royal Society's proposal, and soon offered H.M.S. *Challenger* for a voyage that was to rank with those of Columbus, Magellan, and Cook, especially scientifically. Wyville Thomson was named director of the civilian scientific staff, and he was joined by H. N. Moseley, a naturalist from Oxford, and by two faculty members from Edinburgh, J. Y.

Buchanan, a chemist, and John Murray, then a naturalist and later to rank with Thomson as a pioneering oceanographer.

The *Challenger* was a warship of the old sailing class with a flush deck, equipped with auxiliary engines of more than a thousand horsepower. The conversion from a corvette to a seagoing laboratory was a time-consuming operation, but it went forward smoothly. Thomson himself was a fine planner and organizer, and Captain George S. Nares was a first-rate seaman and an experienced survey officer.

The removal of sixteen of the *Challenger*'s eighteen large guns left the main deck mostly free for the scientific work to come. Cabins were partitioned off and cleared, and efficient biological and chemical laboratories were set up. These were equipped with a full array of instruments, apparatus, and reagents. The fore-cabin was fitted out with work tables and writing tables, and rows of bookcases were well stocked with scientific reference works, and also with old favorites in a lighter vein for recreational reading.

The *Challenger* returned to anchor at Spithead on the evening of May 24, 1876. She had sounded, dredged, and trawled at 362 stations, and had brought back records and specimens that were to take specialists twenty years to describe and classify.

Throughout the long, long voyage Wyville Thomson maintained the best of spirits and the highest enthusiasm. With some wonder, H. N. Mosely gave this picture:

"At first, when the dredge came up every man and boy in the ship who could possibly slip away, crowded round it, to see what had been fished up. Gradually, as the novelty of the thing wore off, the crowd became smaller and smaller, until at last only the scientific staff, and usually Staff Surgeon Crosbie, and perhaps one or two other officers beside the one on duty, awaited the arrival of the net on the dredging bridge, and as the same tedious animals kept appearing from the depths in all parts of the world, the ardour of the scientific staff, even, abated somewhat, and on some occasions the members were not all present at the critical moment, especially when this occurred in the middle of dinner-time, as it had an unfortunate propensity of doing. It is possible even for a naturalist to get weary even of deep-sea dredging. Sir Wyville Thomson's enthusiasm never flagged, and I do not think he ever missed the arrival of the net at the surface."

With this picture of Wyville Thomson we have something of the character of the man. Surely nothing would have pleased him more than to see the voluminous results of the *Challenger* expedi-

Zoological laboratory aboard the Challenger.

Chemical laboratory aboard the Challenger.

The Challenger *off Kerguelen, a desolate island in the South Indian Ocean at about latitude 50° south. From a sketch by J. J. Wild, February 1874.*

tion into final publication—"a full account," as the Royal Society had said, "of the voyage and its scientific results." But that was not to be. Thomson's health broke down soon after the voyage ended, and he died in 1882, a few days after his fifty-second birthday. In his closing years he did manage to write a two-volume account of the Atlantic leg of the *Challenger* voyage, and he had planned companion volumes on the Pacific. But his long years of "work [were] done with the regulation expenditure of tissue; the strain both mental and physical was long and severe, and it has told a great deal upon all of us." He did not live to attempt the story of the Pacific.

John Murray, who was a colleague of Thomson's at Edinburgh and who accompanied him throughout the *Challenger* voyage, saw the results through to final publication.

But we must not leave Wyville Thomson without mention of the great sense of fun that lightened even his most arduous days. With what mirth he followed the antics of Robert, the dilapidated gray parrot that Moseley had acquired in Madeira. As Moseley reported it, "Robert survived all the extremes of the heat and cold of the voyage and perils of all kinds, from heavy tumbles, driving gales of wind, and the falling about of books and furniture. He had one of his legs crippled, and his feathers never grew properly, but he was perfectly happy, and from his perch, which was one of the wardroom hat-pegs, he talked away and amused us during the whole voyage. His great triumph, constantly repeated, was 'What! Two thousand fathoms and no bottom? Ah, Doctor Carpenter, F.R.S.' "

And Thomson was highly amused when Moseley wrote his name with his finger on the surface of a giant *Pyrosoma*[1] (over four feet long) "as it lay in a tub at night, and [his] name came out in a few seconds in letters of fire!"

---

[1] The giant *Pyrosoma* is not a single animal, but rather a colonial organization of many small animals of the tunicate suborder. When a colony is disturbed it displays a brilliant bioluminescence.

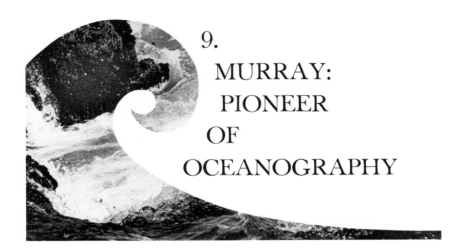

9.
MURRAY: PIONEER OF OCEANOGRAPHY

Nearly a quarter of a century passed between the departure of the *Challenger* from Plymouth and the historic day when the fiftieth volume of the *Challenger* Reports came off the press. The man who made the greatest contribution to that work was John Murray. He was the master unifier. He took strands from all the *Challenger* discoveries and wove them into one fabric—the science of oceanography. He spun contributions from botany, zoology, and physiology; from chemistry, physics, mechanics, and meteorology; from geology, paleontology, and geography, into a new way for man to look at his world, its oceans, and his own relationship to them.

Soon, major universities established professorships in oceanography. One of the first of these was at the University of Liverpool. Appropriately, it was occupied by a former student and intimate associate of John Murray's—William A. Herdman—who held the chair with distinction for many years. In his retirement Herdman wrote a book called *Founders of Oceanography and Their Work*

(1923), which, because of the close association of the two men, contains a uniquely revealing chapter on Murray's life and work. I know no better way to tell the story than to quote this chapter, and I do so now, with the kind permission of the publishers, Edward Arnold and Company, of London.

JOHN MURRAY: PIONEER OF OCEANOGRAPHY

[In this sketch of Sir John Murray] I shall begin with some account of the man, his surroundings and the conditions in which he did his work, and then deal with some of his contributions to oceanography. Murray's period was absolutely continuous with that of Sir Wyville Thomson, and in fact overlapped it; so that, as we shall see, it fell to Murray to continue and complete the work of Thomson, in addition to undertaking other more recent investigations. While Sir Wyville Thomson's name will always be remembered as the leader of the "Challenger" expedition, Sir John Murray will be known in the history of science as the naturalist who brought to a successful issue the investigation of the enormous collections and the publication of the scientific results of that memorable voyage: these two Scots share the honour of having guided the destinies of what is still the greatest oceanographic expedition of all time. [As of 1923.]

John Murray, although a typical Scot in all his ways, was born in Canada—at Coburg, Ontario, on March 3, 1841. But he was of Scottish descent, and returned in early life to maternal relatives in Scotland to complete his education. The lives of our three pioneers just occupied a century (1815 to 1914), and to some extent overlapped. Forbes was only fifteen years senior to Wyville Thomson, and Thomson eleven years senior to Murray. While John Murray was still a school-boy in Upper Canada, Forbes was running his brief meteoric career as professor in Edinburgh, and Wyville Thomson was a young lecturer on the natural sciences in Ireland. Curiously enough, all three went through unusually extended courses as students of medicine and science at the University of Edinburgh, and not one of them took a degree. Forbes was a genius who neglected his work and frankly "funked" his examinations when the time came. In Thomson's case ill-health, fortunately for science, stopped his proposed career in medicine; while Murray despised examinations and degrees, and probably never proposed to take them. He studied a subject because he wanted to know it, and in that spirit he ranged widely over the

Faculties of his university. When I was a student and young graduate I used to hear him denounce in vigorous language all examinations and other formal tests of knowledge, and yet, late in life, there was probably no man of his time who had so many honorary degrees and titles conferred upon him by the universities and learned academies of Europe and America.

After returning to Scotland as a boy in his teens, he lived for some time with a grandfather at Bridge of Allan, and attended the High School at Stirling. During this time he seems to have been most interested in physical science, and especially electricity. He established some electrical apparatus at his home, and in an address to his old school, in 1899, he gives an amusing account of some of the results of his experiments with a large induction coil, such as the following: "On another occasion, several companions arrived from Stirling to see my experiments; they had with them five dogs, one of them being 'Mysie,' a large dog belonging to Sir John Hay, and I had a large Newfoundland called 'Max.' We resolved to give the dogs a shock. They were duly arranged in the room, and the circuit was completed by bringing the noses of the two largest dogs together. Pandemonium was the result. Each dog believed that he had been bitten by the other. They fought, chairs and tables were overturned, and much of the apparatus was broken. In the future, I was requested to turn my attention to the observational sciences of botany, zoology, and geology."

He then spent some years, in the sixties, at the University of Edinburgh, where he was known as a "chronic" student, working at the subjects in which he was interested without following any definite course. Amongst the professors under whom he studied at that time, and who became his close friends in later life, were P. G. Tait in physics, Crum-Brown in chemistry, Turner in anatomy and Archibald Geikie in geology. A decade or so later, after the return of the "Challenger" expedition, he became once more a student at the University of Edinburgh, and that was when I had the good fortune first to meet him.

In 1868 he visited Spitzbergen and Jan Mayen and other parts of the Arctic regions on board a Peterhead whaler, on which, on the strength of having once been a medical student, he was shipped as surgeon. This voyage of seven months probably did much to confirm that interest in the phenomena and problems of the ocean which had been first aroused on his passage home from Canada, ten years before. This interest was doubtless further stimulated during the immediately following years by the epoch-making results of the pioneer deep-sea expeditions in the "Lightning" and "Porcupine," then exploring, under the direction of Wyville Thomson, Carpenter, and Gwyn Jeffreys, the Atlantic coasts of Europe. And then, fortunately, in 1870, Wyville Thomson was appointed professor at Edinburgh, which now became the centre of the negotiations and arrangements with the Admiralty and the Royal Society that led eventually, in 1872, to the equipment and despatch of our great British Deep-sea Exploring Expedition.

It was only an odd chance that led to Murray's connection with the "Challenger." The scientific staff had already been definitely appointed when, at the last moment, one of the assistant naturalists dropped out, and, mainly on the strong recommendation of Professor Tait, in whose laboratory Murray was at the time working, Sir Wyville Thomson offered him the vacant post—surely one of the best examples in the history of science of the right man being chosen to fill a post.

In addition to taking his part in the general work of the expedition, Murray devoted special attention to three subjects of primary importance to the science of the sea, viz., the plankton or floating life of the oceans, the deposits forming on the sea bottoms, and the origin and mode of formation of coral reefs and islands. It was characteristic of his broad and sympathetic outlook on nature that, in place of working at the speciography and anatomy of some group of organisms, however novel, interesting, and attractive to the naturalist the deep-sea organisms might seem to be, he took up wide-reaching general problems with economic and geological as well as biological applications. Amongst the preliminary reports sent home during the course of the expedition, and published in the *Proceedings of the Royal Society,* we find those by John Murray, written from Valparaiso, December 9, 1875, dealing with (1) Ocean Deposits, (2) Surface Organisms and their relation to Oceanic Deposits, and (3) Vertebrata (mainly Fishes), which, though superseded by the later work of himself and others, are still of great historic interest. In that preliminary account of the Ocean Deposits we find Murray's first classification into (1) Shore deposits, (2) Globigerina ooze, (3) Radiolarian ooze, (4) Diatomaceous ooze, and (5) Red and Grey Clays, which has been adopted with little or no change in all succeeding works; and, in his report on the surface organisms, we find the first figures of the living *Hastigerina, Pyrocystis,* and the remarkable deep-water Radiolaria known as "Challengerida."

Each of the three main lines of investigation—deposits, plankton, and coral reefs—which Murray undertook on board the "Challenger" has been most fruitful of results both in his own hand and those of others. His plankton work has led on to those modern plankton researches which are closely bound up with the scientific investigation of our sea-fisheries. His observations on coral reefs, in conjunction with the "Challenger" results as to depths of the ocean and the presence of submarine volcanic elevations, resulted in his new and most original theory as to the formation of "atolls," which removed certain difficulties that had long been felt by zoologists and geologists alike to stand in the way of the universal acceptance of Darwin's well-known theory of coral reefs and islands.

His work on the deposits accumulating on the floor of the ocean resulted, after years of study in the laboratory as well as in the field, in collaboration with the Abbé Renard of the Brussels Museum, afterwards Professor at Ghent, in the production of the monumental *Deep-sea Deposits* volume, one of the "Challenger" reports, which first revealed to the scientific world the detailed nature and distribution of the varied submarine deposits of the globe and their relation to the rocks forming the crust of the earth.

These studies led, moreover, to one of the romances of science which deeply influenced Murray's future life and work. In accumulating material from all parts of the world and all deep-sea exploring expeditions for comparison with the "Challenger" series, some ten years later, Murray found that a sample of rock from Christmas Island, in the Indian Ocean, which had been sent to him by Commander (now Admiral) Aldrich, of H.M.S. "Egeria," was composed of a valuable phosphatic deposit.

Murray's interest in this rock was at first solely in relation to the "Challenger" deposits and its possible bearing on his coral-reef theory; but he soon realized its economic as well as scientific interest, and was convinced that the island would be of value to the nation. After overcoming many difficulties, he induced the British Government to annex this lonely, uninhabited volcanic island, and to give a concession to work the deposits to a company which he formed. He sent out scientific investigations to study and report on the products, and the results have been highly successful on both the scientific and the commercial sides. Sir John Murray visited Christmas Island himself on several occasions, he had roads cleared, a railway constructed, waterworks established, piers built, and the necessary buildings erected. In fact, the lonely island was colonized by about 1,500 inhabitants, and flourishing plantations of various kinds were established in addition to the mining of the phosphatic deposits. Murray was able to show that some years before the war [World War I] the British Treasury had already received in royalties and taxes from the island considerably more than the total cost of the "Challenger" expedition. This is one of those cases where a purely scientific investigation has led directly to great wealth—wealth, it may be added, which in this case has been used to a great extent for the advancement of science.

In the case of Sir John Murray . . . I am writing of a man who made a strong personal impression as one of my teachers in science at Edinburgh some forty-five years ago. It is not from one's formal instructors alone that one learns. Murray was never on the teaching staff of the university; but a few of us (generally Major-General Sir David Bruce, now of the Lister Institute, Professor Noel-Paton, now of Glasgow, and myself) who were then, in the late seventies, young students of science, and were privileged to have the run of the "Challenger" Office, learned more of practical Natural History from John Murray than we did from many university lectures.

This was in the few years following on the return of the "Challenger" expedition in 1876, and the vast collections of all kinds brought back from the seas and remote islands were being classified and sorted out into groups for further examination in a house near the university, known as the "Challenger" Office. Murray, as First Assistant on the Staff, had charge of the office and the collections, and welcomed a few eager young workers who were willing to devote free afternoons to helping in the multifarious work always in progress.

There we first made acquaintance with the celebrated new deep-sea "oozes," learnt to distinguish them under the microscope, and how to demonstrate the silicious Radiolaria hidden in the calcareous Globigerina ooze; and there we first saw such wonders of the deep as *Holopus* and *Cephalodiscus,* and the extraordinary new abyssal Holothurians, afterwards known as Elasipoda. These —now the commonplaces of marine biology—were then revelations, and those of us who witnessed the discoveries in-the-making will always associate them with "Challenger Murray" as the archmagician of the laboratory—a sort of modern scientific alchemist, bringing mysterious unknown things out of store-bottles, and then showing us how to demonstrate their true nature. I am afraid that we who are trying to inspire students with the sacred fire at the present day have no such wonders to show as these first-fruits in the early days of deep-sea research. Then between times, while waiting for a reaction, or after work, Murray would tell us stories of the great expedition—how the first living *Globigerina* (*Hastigerina murrayi*), seen in all the glory of its vesicular protoplasm expanded far beyond its tiny shell, was picked up in a teaspoon from a small boat during a dead calm in mid-ocean; and how the naval officers wrote their names with their fingers with letters of fire on the phosphorescing giant *Pyrosoma* (over four feet long) as it lay on the deck at night; how they "iced" their champagne in the tropics by plunging the bottles into the trawlful of ooze just brought up from the abyss, and still retaining its abyssal low temperature; and, finally, he would sing us a most amusing song—we never knew whether he had invented it or not—about a Chinaman eating a little white dog.

A few years later, after Sir Wyville Thomson's death in 1882, Murray had supreme control of both the collections and the editing of the reports; and of the "Office," by that time moved to more commodious quarters at 32 Queen Street, which was the scene of his labours for many years, and where I for a time held the post of "Assistant-Naturalist," and saw Murray practically every day.

When I first knew John Murray, although he was an older man, we were really in one respect fellow-students, as we attended together Professor Archibald Geikie's course on geology. One very pleasant and not the least instructive part of the course at that time was the series of geological walks personally conducted by the professor, not merely Saturday walks in the neighborhood of Edinburgh, but also longer expeditions of a week or ten days at the end of the session, to localities of special geological interest farther afield, such as the Highlands, or the Island of Arran. I well remember one such long excursion to the Grampian and the Cairngorm Mountains and Speyside, when he had, as somewhat senior members of the party—in addition to Professor Geikie—Dr. Benjamin Peach and Dr. John Horne of the Geological Survey, Dr. Aitken of the University Chemical Department, Joseph Thomson the African explorer, and John Murray of the "Challenger." The rest of us were ordinary students of science, and all will realize how we enjoyed and profited by the conversation of these senior men, how we dogged their steps and hung upon their every word. All who ever met John Murray will readily understand that in the frequent discussions that took place between these geologists and chemists, he always took a leading and forcible part—he was nothing if not original in his views and vigorous in his language.

The reader need not think that all this had nothing to do with oceanography. It was very much otherwise. These were all Edinburgh men deeply interested in the "Challenger" results. On the long tramps there were hot discussions, and wherever Murray was he was apt sooner or later to bring a discussion round to some fundamental problem of the ocean or the deposits forming on its floor, or to illustrate an argument by something he once saw in the Pacific, or the Antarctic—or elsewhere. And, moreover, on the tops of these ancient mountains of Scotland we could, and did, consider the changes of continents and the supposed permanence of ocean basins. I, for one, then came to realize that geology has a close bearing on oceanography; and I suspect that it was on occasions like these, in keen discussions with geologists and chemists, that Murray formulated some of the theories as to past history of land and sea that he afterwards published in the *Summary* volumes of the "Challenger" series.

Murray's first paper on his theory of coral reefs was read before the Royal Society of Edinburgh on April 5, 1880, and was published in the *Proceedings,* vol. x., p. 505. I well remember the occasion, and also the rehearsal which took place some days before in Sir Wyville Thomson's house of Bonsyde, when Murray read his MS. to a small but highly critical audience, consisting of Sir Wyville Thomson, Sir William Turner, and myself. For months before I had daily seen Murray preparing the paper in a large room at the "Challenger" Office, sitting at his notes in the centre of a multitude of charts showing all the reefs and coral islands of tropical seas—some of the charts spread out on tables, others carpeting the floor or stacked in piles and rolls—while he measured and drew sections of the contours so as to see which reefs supported his views and which presented difficulties. His coral-reef theory was the direct outcome of his "Challenger" work. The soundings had revealed the presence of volcanic elevations, and the distribution of the calcareous deposits showed how these might contribute to build up suitable platforms as the foundation of reefs which might grow to the surface independent of sunken lands such as Darwin's theory had required. It may be said that Murray demolished the supposed need of vast oceanic subsidence, which had been felt to be a difficulty by many geologists, and showed that all types of coral reef could be accounted for without subsidence, and even in some cases with elevation of land.

Some of Murray's friends were disappointed that his theory did not receive more serious and more immediate attention, and the then Duke of Argyll wrote a couple of articles with somewhat sensational titles—"A Great Lesson," in the *Nineteenth Century* . . . and "A Conspiracy of Silence," in *Nature* . . . which gave rise to answers from some of the leading men of science of the day, Huxley, Bonney, and Judd. Murray went on his way undisturbed, collecting further evidence and publishing at intervals further papers dealing with one or another part of the large subject—such as his paper on the structure and origin of coral reefs in the *Proceedings* of the Royal Institution for 1888, his account of the Balfour Shoal in the Coral Sea (1897), a submarine elevation in process of being built up by calcareous deposits, his "Distribution of Pelagic Foraminifera at the surface and on the floor of the Ocean" (1897), and a series of reports upon bottom deposits from the "Blake" (1885) and many other expeditions.

Later on (1896–8) Murray took a lively interest in the investigation, by a Committee of the British Association and the Royal Society, of a selected typical case, the atoll of Funafuti, one of the Ellice Group, in the South Pacific. A first expedition was sent out from this country under Professor Solas, and then two others from Australia, under Professor Edgeworth David, of Sydney, and borings were eventually obtained reaching an extreme depth of over 1,100 feet. The core was brought home and subjected to detailed microscopic examination, with the extraordinary result that the supporters of both rival theories find that it can be interpreted so as to support their views. The Funafuti boring cannot be said to have settled the matter. I believe the verdict at the present time of most zoologists and geologists would be that whereas Darwin's beautiful theory would certainly hold good for coral reefs growing on a sinking area, Murray's explanation, based upon other observations and ascertained facts, probably applies to many of the "atolls" and "barrier reefs" of tropical seas.

But I have been led on to these more recent times by his paper of 1880. Let us now return to his work at the "Challenger" Office. During the last couple of years of Sir Wyville Thomson's life, when he was more or less of an invalid, Mr. John Murray (as he then was) came gradually to take over more and more complete charge of affairs at the "Challenger" Office, including the distribution of the groups of animals to specialists and the editing of the volumes of reports. It was very fortunate for zoological science that such a man was on the staff, ready to take up and carry out to a successful issue the work that Sir Wyville Thomson was no longer able to continue. Murray brought to the task a complete knowledge of all that had to be done and how best to do it, along with an extraordinary amount of zeal and energy. During the years that followed, until the completion of the work, he seemed to be doing several men's work. He was in constant communication, both by correspondence and personal visits, with the authors of reports in various parts of Europe and America; he had frequent dealings with the Government departments concerned in the production of the work; and all the time he was also himself investigating some of the great general problems of oceanography. It is difficult to imagine that any other man than John Murray could have carried through all this mass of detailed and difficult work and have produced the fifty thick quarto volumes within twenty years of the return of the expedition. About five of these large volumes are the result of Murray's own work. Along with Staff-Commander T. H. Tizard, the late Professor H. N. Moseley, and Mr. J. Y. Buchanan, he drew up the general *Narrative of the Expedition;* along with the late Professor Renard he wrote the very important report upon the *Deep-sea Deposits* (1891), generally recognized as the authoritative work on the subject; and finally, at the conclusion of the series, he produced two volumes entitled *Summary of the Results* (1895), which give an elaborate historical account of our knowledge of the sea and the development of the science of oceanography from the earliest times to the present day, and also, in addition to complete lists of all the organisms at all the "Challenger" stations, includes a discussion of many important matters, geological as well as biological, relating to the origin of the present configuration of land and water and of the distribution of the marine fauna and flora of the globe.

It was characteristic of him to put forward, especially in these *Summary* volumes, views which were novel and even daring, which he believed he had evidence to support, but a less courageous man might have kept back or expressed more cautiously. He always had the courage of his convictions. He admitted that he sometimes made mistakes, but he held that the man who never made a mistake never made anything else. That was one of his *obiter dicta* which were flying about the "Challenger" Office, and stuck in my impressionable youth. Let me quote here a passage from one of his many letters that I have, and which refers to the kind of views he afterwards published in his *Summary*. It is dated September 13,

1894, and is evidently in answer to some question I had asked as to his views on the past history of life in the sea.

". . . I gave two papers to the R.S.E. and also said something about distribution at the British Association, but I have not yet published anything. I am now considering whether or not I will add a chapter to the last 'Challenger' volume, giving my views.

"I believe the continental areas are very permanent, and for instance Africa has separated marine faunas and floras longer than the time when there was a very nearly similar fauna at both poles. However, the faunas of the sea are now arranged more according to zones of temperature than by land barriers. The tropics extend polewards as we go down in the geological formations till just before the Chalk there was a universally warm sea—from equator to poles and from top to bottom—say 80° F. Coral reefs once flourished at the poles. These have now been driven to equatorial regions where the temperature has remained nearly the above. The animals which in the universal warm sea came to live in the mud at a little depth, remained behind when cooling of the poles commenced. These animals without pelagic free-swimming larvae also descended to the deep sea as the waters cooled. When the sea was all 70° or 80° F. the deep sea was not inhabited. Polar animals and deep-sea animals have all a direct development (so also fresh-water animals, also derived from the deeper part of the shore estuarine universal fauna).

"It is nonsense to suppose that while the earth was developing the sun has always been the same as now. It has been contracting. In Chalk times it had a diameter seen from the earth equal to an angle of 10° in the heavens. This would give all the heat and *light* that is necessary for a great Carboniferous forest at the poles.

"You can tell me how much of this is d——d nonsense.

"Yours sincerely, John Murray.
"Fresh water fauna is much more archaic than deep-sea."

The following, from his little book *The Ocean* (p. 226), is a good example of Murray's bold speculations: "We look back on a past when the crust of the earth was in a molten condition with a temperature of 400° F., when what is now the water of the ocean existed as water vapour in the atmosphere. We can imagine a future when the waters of the ocean will, because of the low temperature, have become solid rock, and over this will roll an ocean of liquid air about forty feet in depth."

One of the theories which he supported, and which is not now generally accepted, although he believed he had much evidence in favour of it from the "Challenger" results, was the theory of "Bipolarity," viz., that identical organisms were found in Arctic and Antarctic seas and not in intermediate waters, and that they represented the original marine fauna which at some earlier period of the earth's history inhabited all the oceans. This bipolarity hypothesis has been vigorously controverted, and, like some other theories in science which have had to be abandoned, was most useful in its day as giving rise to much new investigation. A good deal of evidence against Murray's views on bipolarity has been accumulated as the result of recent Antarctic expeditions.

But whether all his views are accepted or not, they are all very stimulating and useful, and have given rise to much investigation and discussion in the history of oceanography. His five great volumes are a notable monument to his memory. They and the other "Challenger" reports which he edited record collectively the greatest advance in the knowledge of our planet since the great geographical discoveries of the fifteenth and sixteenth centuries.

[In 1880–2 William B. Carpenter] and Wyville Thomson, during the preliminary investigations in the "Lightning" and "Porcupine," had found that the Faroe Channel was divided into two regions—a "cold" and a "warm" area. The temperature of the water to a depth of 200 fathoms is much the same in the two areas; but in the cold area to the N.E. the temperature is about 34°F. at 250 fathoms, and about 30° at the bottom at 640 fathoms, while in the warm area, which stretches S.W. from the line of demarcation, the temperature is 47°F. at 250 fathoms, and 42° at the bottom in 600 fathoms. A consideration of the "Challenger" temperatures led to the conclusion that the cold and warm areas of the Faroe Channel must be separated by a very considerable submarine ridge rising to within 200 or 300 fathoms of the surface. Sir Wyville Thomson induced the Admiralty to give the use of a surveying vessel for a few weeks for the purpose of sounding the Faroe Channel with a view of testing this opinion. That was the origin of the "Knight-Errant" expedition in the summer of 1880, conducted by Captain Tizard, R.N., and Mr. John Murray, under the general direction of Sir Wyville Thomson, who remained at Stornoway, in the Outer Hebrides, during the four traverses of the region in question. The results showed that a ridge rising to within 300 fathoms of the surface runs from the N.W. of

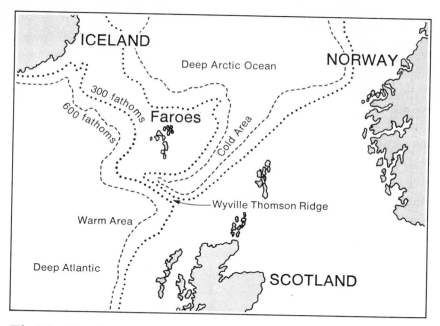

The Wyville Thomson Ridge in the Faroe Channel.

Scotland by the island of N. Rona to the southern end of the Faroe fishing-bank.

This was followed, after the death of Sir Wyville Thomson, by a further expedition in H.M.S. "Triton," in the summer of 1882, again under Murray and Tizard, which was very fruitful of zoological results. The discovery of two very different assemblages of animals living on the two sides of the Wyville Thomson ridge—Arctic forms to the North and Atlantic forms to the South—gives us a notable example of the effect of the environment on the distribution of marine forms of life. The results of the "Triton" expedition, written by a number of specialists, were published in the *Trans. Roy. Soc. Edin.* during the next few years, and attracted much attention to the subject.

Dr. Johan Hjort, the Norwegian oceanographer, referring some thirty years later to these expeditions, said (*The Depths of the Ocean,* 1912, p. 661): "In the history of oceanic research possibly nothing has contributed so much to the awakening of this interest as the discovery of entirely different animal communities living on either side of the Wyville Thomson Ridge. Atlantic forms occur to the south and Arctic forms to the north of the ridge, corresponding to the very different thermal conditions on either side."

During these few years after the "Triton" expedition, and when, in consequence of Sir Wyville Thomson's death, he was given complete charge of the "Challenger" Office, Murray came to occupy a more and more prominent position in the scientific world of the North. When we remember that his earlier fellow-workers and associates at the university were such men as Robertson Smith the theologian, Dittmar the chemist, Sir John Jackson the great contractor, and Robert Louis Stevenson; and his later friends, after the return of the "Challenger," were such men as Agassiz, Turner, Crum-Brown, Tait, Renard, Haeckel, Geikie, Blackie, Masson, Buchan, and Lord McLaren, we can understand the stimulating atmosphere he lived and worked in, and to which he doubtless contributed as much as he received.

We now come to a period of great local scientific activity, when Murray exercised a notable influence in the university scientific circle and took a leading part in every new movement. He was a prominent member of the Royal Society of Edinburgh, and of the Scottish Meteorological and Geographical Societies; he helped to establish the Observatory on the summit of Ben Nevis; and in 1884, along with his friend, Robert Irvine, of Caroline Park, on the shores of the Firth of Forth, he acquired the lease of an old sandstone quarry near Granton, into which the sea had burst some thirty years before, drowning the quarry and leaving it as a land-locked sheet of sheltered deep water which rose and fell with every tide. Here he moored a large canal barge, upon which he had built a wooden house, divided into chemical and biological laboratories, and which, for obvious reasons, he named "The Ark." Two little Norwegian skiffs were attached to "The Ark," one for the chemists and the other for the biologists, and on the opening day Dr. Hugh Robert Mill and I were invited to name them. He called his "The Asymptote," and I named the other "Appendicularia." Murray ridiculed our pretentious names, and said that in a few days the one would probably be called "the Simmie," or "the Tottie," and the other, "Dick."

This floating biological station, after some years' work at Granton, was towed through the Forth and the Clyde Canal to Millport, on the Cumbrae Island, and there it was beached and became an annex of the Millport biological station. During the period when "The Ark" was at Granton, and later, Murray and Irvine turned out a good deal of joint work on the chemistry of the secretion of carbonate of lime by marine organisms, on the solution of carbonate of lime by the carbon dioxide in sea-water, and on the chemical changes taking place in muds and other deposits on the sea bottom.

But his chief scientific work at this time and for years afterwards was the joint investigation at the "Challenger" Office of the enormous series of deposits (said to be over 12,000) which he and the Abbé Renard had accumulated from many expeditions and all seas. When one entered the little laboratory on the top floor of 32 Queen Street, after penetrating the dense cloud of tobacco smoke, the first thing one heard, rather than saw, was John Murray issuing some order or announcing some result; the next was the figure of the portly Abbé waving a courteous greeting with his perpetual cigar. Then there were the two assistants, Mr. F. Pearcey, who had himself, as a boy, taken part in the great expedition, and had been retained as assistant curator of the collections at the "Challenger" Office; and Mr. James Chumley, the secretary. Murray and Renard were hard at work at the microscope or at chemical reactions in test-tubes over Bunsen burners, Pearcey was preparing fresh samples to be examined, and Chumley was noting down results. There has probably never been in recent years such a small laboratory, so poorly equipped, which has turned out such epoch-making results. Everything absolutely essential was there, but nothing in the least extravagant. The place looked, with its plain boards and deal tables and sinks, more like an overcrowded scullery than an oceanographic laboratory.

But even in his busiest years at the "Challenger" Office Murray never gave up wholly his work at sea. He was a good hand at "roughing it" and making the best of circumstances, and no one could have had a greater appreciation of the open-air life. The practical work that he did, more or less periodically all the year round, on the west coast of Scotland, from his little yacht "Medusa," is a good example of careful planning and resolute carrying out.

It seems that while working at the results of the "Challenger" and other deep-sea expeditions, it occurred to Murray that for the purpose of comparison a detailed examination of the physical and biological conditions in the fjord-like sea-lochs of the West of Scotland might yield valuable information. He accordingly built a small steam-yacht of about 38 tons, called the "Medusa," fitted up with all the necessary apparatus for dredging and trawling and for taking deep-sea temperatures and other hydrographic observations. This little vessel was, in fact, fully equipped for oceanographical investigations in the neighborhood of land, and during the years 1884 to 1892 she was almost continuously engaged in exploring the deep sea-lochs of the Western Highlands. Various younger scientific men, such as Dr. W. E. Hoyle and Dr. H. R. Mill, were associated with Murray in this work; considerable collections were made, some of which are now in the British Museum, and many scientific papers contributed to various journals have resulted from the periodic cruises of the "Medusa." One of the most notable of these is H. R. Mill's detailed description of the oceanographic characters of the Clyde sea-area (1891–4). Another result was the discovery in the deeper waters of Loch Etive and Upper Loch Fyne of the remnants of an Arctic fauna—"boreal outliers" of Edward Forbes.

From time to time during these researches in the sea-lochs, the "Medusa" penetrated to the fresh-water lochs, such as Loch Lochie and Loch Ness, which are united by the Caledonian Canal, and Murray was greatly impressed by the differences in the physical and biological conditions between the salt and the fresh-water lochs. This observation seems to have led to another of Murray's scientific activities, namely, the bathymetrical survey of the fresh-water lochs of Scotland, undertaken between the years 1897 and 1909. It was already known that, like some of the salt-water fjords outside, certain of these fresh-water lochs are of surprising depth. For example, 175 fathoms had been recorded by Buchanan in Loch Morar, and Murray, subsequently running a line of soundings along this loch, found at one spot a depth of 180 fathoms. No such depth is found in the sea outside on the continental shelf.

The survey was undertaken at first in collaboration with his young friend, Mr. Frederick P. Pullar, who was drowned in a gallant attempt to save the lives of others in a skating accident on Loch Airthrey in 1901. The results of the Lake Survey were published in a series of six volumes (Edinburgh, 1910), edited by Sir John Murray and Mr. Lawrence Pullar, and dedicated to the memory of Mr. F. P. Pullar, who had done much to initiate and promote the investigation in its earlier stages.

The work dealt with the determination of the depths of the lakes and the general form of the basins they occupy, along with observations in other branches of limnography from the topographical, geological, physical, chemical, and biological points of view. Some important novel investigations, such as those on the temperature seiche [sudden oscillation] and variations in the viscosity of the water with temperature, help to throw light on some oceanographical problems. In fact, the whole investigation, comprising 60,000 soundings taken in 562 lakes, resulted in very substantial contributions to knowledge, and is probably the most complete account of the depths and other physical features of lakes that has been published in any country.

It cannot be said that Murray ever finished his work on the west coast of Scotland, and I have evidence in a letter that he wrote to me late in life that he still thought of returning to the work. The passage is worth quoting, both for its scientific interest and for the kindly consideration that it shows. It is dated May 20, 1913, less than a year before his death:

". . . I am seriously thinking of overhauling all the 'Medusa' work on the west coast, and repeating a lot of those old observations for two years or more; then publishing a book on the lochs of the west coast. Would that in any way interfere with your work? I am being pressed by the Clyde people to do something of the kind.

"Could I afford it at present, I would be off to the Pacific in a Diesel-engined ship!!"

During the years when he was working at the "Challenger" results and subsequently Murray published many papers in the *Geographical Journal* and in the *Scottish Geographical Magazine* and elsewhere, which dealt with world-wide questions in oceanography or in physical geography, such as the annual rainfall of the globe and its relation to the discharge of rivers, the effects of winds on the distribution of temperature in lochs, the annual range of temperature in the surface waters of the ocean, and the temperature of the floor of the ocean, on the height of the land and the depth of the ocean (1888), and on depths, temperatures, and marine deposits of the South Pacific Ocean (1906).

In 1897 Dr. John Murray (as he then was) formally opened the present Biological Station at Millport and the associated Robertson Museum, and delivered an address on the marine biology of the Clyde district. He continued to take a lively interest in the affairs of the West Coast Biological Station, and frequently looked in there with scientific friends when on his cruises in the "Medusa." I recollect for example, an occasion when, after dredging in Loch Fyne, we ran to Millport for the night, and the party included Canon Norman, old Dr. David Robertson, Professor Haeckel, and Mr. Isaac Thomson. He frequently had foreign men of science as his guests, and was, I think, especially friendly with the Scandinavians, such as Nansen, Hjort, Otto Pettersson the

Swede, and C. G. Joh. Petersen the Dane.

Murray's oceanographic work was not limited to any particular region or special series of problems, but was world-wide, both in extent and in subject-matter. He was a great traveller, and had probably personally explored more of the oceanic waters of the globe than any other man. He had ranged from Spitzbergen in the North to the Antarctic Ice-barrier, dredging, trawling, tow-netting, and sampling the waters and bottom deposits in every possible way. Even when travelling as an ordinary passenger on a liner, he would engage emigrants in the steerage to pump water daily from the sea through his silk nets, or would arrange with a bath steward to let the sea-water tap run through his net day and night in order that he might have living plankton to examine.

Murray was not only an investigator of special problems, but we owe to him much synthetic work, in which he gathered together the results of many observations and put them in the form of short conclusions or statistical statements. Some of these were published in the form of useful maps and charts, such, for example, as the map showing the 57 "deeps," or parts of ocean in which soundings of over 3,000 fathoms have been obtained. Most of these deeps (32) are in the Pacific, including the deepest soundings of all, which extend down to over six English miles.

At the meeting of the British Association held at Ipswich in September, 1895, a meeting of contributors to the "Challenger" reports was held, at which the then President of the Zoological Section (W. A. Herdman) presided, and about fifty biologists or oceanographers either attended or wrote expressing their concurrence in the objects of the meeting. It was then proposed and resolved "that this meeting of those who have taken part in the production of the 'Challenger' reports agrees to signalize the completion of the series by offering congratulations in some appropriate form to Dr. John Murray." Eventually this congratulatory offering took the form of an address in an album, containing the portraits and autographs of all the "Challenger" workers, with an illuminated cover and dedicatory design by Walter Crane. This book was afterward reproduced for the contributors in the form of a thin quarto volume, which forms a very interesting record of the completion of the work connected with the "Challenger" expedition.

Dr. Murray himself provided a very pleasing memento of the conclusion of the great work by having a handsome medal designed and struck, an example of which was presented to each of the authors of "Challenger" reports. The medal, in bronze alloy, measures 75 mm. [almost 3 inches] in diameter, and shows on the obverse the head of Minerva encircled by mermaids, a dolphin, and Neptune holding in his left hand the trident, and in his right the naturalist's dredge, with the legend, "Voyage of H.M.S. 'Challenger,' 1872–76"; and on the reverse an armoured knight casting down his gauntlet in challenge to the waters—being the crest of H.M.S. "Challenger"—with the legend, "Report on the scientific results of the 'Challenger' Expedition, 1886–95." The name of the recipient of the medal is engraved on the lower margin.

After Sir Wyville Thomson's death, when Murray came to be recognized by the scientific world as the moving spirit in connection with the "Challenger" work, and especially when the great series of publications was completed, honours of all kinds came pouring in upon him—for which he probably cared little. He was an honorary doctor of many universities, he was awarded the "prix Cuvier" medal by the Paris Academy of Sciences, and he was created K.C.B. in 1898. He gave the Lowell lectures at Boston in 1899, and again in 1911. He was chief British delegate at the International Congress for the Exploration of the Sea, at Stockholm, in 1899. He was President of the Geographical Section of the British Association in the same year; and it is an open secret that he might have been President of the Association had he been able to undertake it. He was approached no less than three times in connection with three different meetings (two of them overseas meetings, at which it was felt that a man of world-wide associations, such as Murray, would be singularly appropriate), but after some hesitation and careful consideration, he felt that circumstances compelled him to decline the honour. Some of his letters to me, from which I quote a few passages, allude to these offers.

This is a letter from Mentone, on April 1, 1904, referring to the first of these occasions:

". . . At first, I said it was impossible to alter our family and other arrangements so as to go to South Africa. . . . To my astonishment my wife seems taken with the idea of going to the Cape, and says it is by no means impossible to alter our arrangements. I've promised to think over the matter for a week. I'll let you know definitely a day or two after I reach Edinburgh.

"I feel that you are predisposed to honour me, but I also feel I have given the Association very little of my attention: others have more claims on the honour. I don't care a bit about it. If I consult my own feelings, I would much rather have nothing to do with it. My wife suggests there may be some question of duty. Perhaps? I had not heard you had taken on the General Secretaryship. . . ."

In a letter from Boston, U.S.A., he writes on March 20, 1911:

". . . On Saturday I received your letter of the 3rd March. By same post had letters from Geikie and Bonney. Had I been at home, I would of course have seen you before sending any reply, but I am not likely to be in England before June.

". . . Tomorrow I deliver the Agassiz address at Harvard. I came over for that address, but have been let in for the Lowell lectures (eight) and addresses here [Boston], Princeton, New York, and Washington. We go to Washington next month. . . .

"During the last two days I've had frequent deliberations with my wife and daughter, who are with me, and the only way out seems to be to decline the nomination. For some time past I have been planning a cruise as far as the Pacific during 1912 and 1913, and I have made a good many business and domestic arrangements with that object in view. It must take place in these years or not at all, and if my health be good I cannot well withdraw.

"I know your enthusiastic nature and your too favourable opinion of my poor labours. I know you like to do me honour. For these reasons I very much regret the nature of the cables I have just sent off to you, Bonney, and Geikie. I am anxious to do anything to assist the progress of oceanography, but I fear my presidentship of the British Association would not do much in that direction. However it is very good and nice of you to say you think it would. I find many enthusiastic young workers here, and I believe there will likely be a ship fitted out for a deep-sea expedition in 1912. They wish to consult me at Washington and New York about this. Townsend is now away in the 'Albatross,' off the Pacific coast. They invited me to go with them, also to go to the Tortugas Station, where some very interesting work is going on. . . ."

This further letter refers to the same occasion. It is from Washington, D.C., April 19, 1911:

". . . I duly received your letter of the 20th. I have not replied at once, especially as I had written to you when I sent off my cable, and I had also cabled and written to Bonney and Geikie. I have not changed my mind about the presidency. I cannot see my way to accept. I am very sorry, for I would willingly do very much to please you and my other friends on the Council. I also believe that some scientific man less known locally would be more agreeable to the Dundee people.

"You will see from the enclosed cutting that they have been doing us much honour here. There was a dinner in our honour last week, about seventy-five scientific men here and their wives. The British Ambassador and his wife were present. Taft accepted, but sent an excuse at the last minute.

"We go to Philadelphia to-morrow to meetings of Philadelphia Academy. Then to New York. Osborn is to have 14 millionaires to hear me at the Museum as to what they should do for the study of the ocean!! May it have some effect!

"On the 29th we start for the West to see rocks and mines in Nevada. We sail from Boston on the 30th May.

"With my very best thanks to you for all your endeavours to honour me, and to cultivate an interest in oceanography."

The following letter of November 12, 1912, refers to the final occasion. He was killed before the meeting in question took place:

". . . I shall not refuse at once. I'll consult with my wife. All the same, I do not think it is the sort of thing for a man of over seventy. I'm very well just now—have been for the past three months shooting over the moors nearly every day! Some people say even that I am a wonder! But who can tell what I'll be like in two years. Men over seventy years are likely to break down, then what a nuisance I would be to everyone!

"I would, of course, appreciate the honour, but honours are not worth much to an old man. The only question would be, a real service to Science, and would it be a duty. At my age it can hardly be a duty. I have no message to give to the world!! I honestly think some young scientific man would do the trick very much better. I'll consider it. I'll be in London, Piccadilly Hotel, the first ten days of December, and could perhaps see you.

"I really very much appreciate your desire to honour me. It is really very good of you. It is not quite out of the possible that I may be in the Pacific in 1914 in a boat of my own. I would have been there now had the cost not been much greater than I, at first, calculated."

At the inauguration of the New Zoological Laboratories of the University of Liverpool in November, 1905, Sir John Murray was one of the honoured guests of the university, and after the formal opening by the Earl of Onslow, Sir John gave a short address upon oceanography, the first lecture to be delivered in the zoology lecture theatre of the university. A few years later, in 1907, the university conferred upon him the honorary degree of Doctor of Science.

We now come to Sir John Murray's last great scientific expedition—a four months' cruise in the North Atlantic, in the summer of 1910—a very notable achievement for a man in his seventieth year. The investigating steamer "Michael Sars" was built by the Norwegian Government in 1900, on the lines of a large high-class trawler of about 226 tons, but specially fitted out for scientific work under the direction of Murray's friend, Dr. Johan Hjort. At Murray's request this vessel was lent, with her crew and equipment, by the Norwegian Government for the North Atlantic cruise, Sir John Murray undertaking to pay all the expenses. The scientific reports on the expedition will be published in a series of volumes by the Bergen Museum; but the more general results have appeared in popular form in a volume entitled *The Depths of the Ocean* by Murray and Hjort, with contributions by several other naturalists, which gives a condensed account of the modern science of oceanography, with special chapters on the latest discoveries, based largely upon the experiences of this North Atlantic cruise taken along with the previous cruises of the "Michael Sars" in the Norwegian seas.

Amongst noteworthy matters that are discussed in this volume we find:

(1) Methods of plankton collecting, including the towing of as many as ten large horizontal nets, at various depths, simultaneously. The pelagic plants collected, either in the nets or by centrifuging the water, are discussed in a notable chapter by Gran.

(2) The "Mud-line," a favourite subject with Murray, as being the great feeding-ground of the ocean. He places it at an average depth of 100 fathoms, on the edge of the "Continental-shelf," at the top of the "Continental-slope," which descends more or less precipitately to the floor of the Atlantic at an average depth of 2,000 fathoms. We know from Murray's careful estimations that, if all the elevations of the globe were filled into depressions, we should have a smooth sphere covered by an ocean 1,450 fathoms deep. The floor of this ocean is the "mean sphere level."

(3) Dr. Helland-Hansen, the physicist on board the "Michael Sars," had devised a new form of photometer, which registered light as far down as 500 fathoms in the Sargasso Sea. At between 800 and 900 fathoms, however, no trace of light was registered

on the photographic plates, even after two hours' exposure. The observations show that light in considerable quantity penetrates to a depth of at least 1,000 metres (547 fathoms), which is much deeper than had been previously supposed. It was shown that the red rays of light are those that disappear first, and the ultra violet are those that penetrate most deeply.

(4) A special study was made on the "Michael Sars" of the characteristic colour of the fishes in various zones of depth. In the superficial layers of the ocean small colourless or transparent forms abound, forming a part of the well-known pelagic fauna. Below this, at an average depth of about 200 fathoms, are found fishes of a silvery and greyish hue, along with red-coloured Crustaceans. At depths of from 500 fathoms downwards black fishes make their appearance, still associated with red Crustaceans and other strongly coloured red, brown, or black Invertebrates. This chapter is illustrated by some beautiful coloured plates of the fishes.

(5) Lastly, the "Michael Sars" got important evidence in support of the view that the fresh-water eel spawns south of the Azores, and that the larvae are carried by currents back to the coasts of North-west Europe.

In 1913 Murray published in the Home University Library a small book of about 250 pages, entitled *The Ocean: A General Account of the Science of the Sea,* which is undoubtedly the most concise and accurate and, so far as is possible within its small compass, complete account that has yet appeared of all that pertains to the scientific investigation of the sea. It is written in simple language for the general reader, and is probably the best introduction to oceanography that can be recommended to the junior student or the intelligent non-specialist inquirer who desires information merely as a matter of general culture. It deals with the history, methods, and instruments of marine research, the depths and physical characters of the ocean, the circulation of the waters, life in the ocean, submarine deposits, and finally the nature and relations of the various "Geospheres" that constitute the globe. Coloured maps and plates illustrate depths, salinities, temperatures, currents, deposits, and many of the characteristic plants and animals of the plankton and of the "oozes." As Murray's final contribution to science it is an appropriate summary of his life-work, and will do much to spread the knowledge of his discoveries and to make his name widely known amongst readers of popular works on science.

If I try now to give a personal impression of John Murray as I remember him in earlier life, I picture him as a short, thick-set, broad-shouldered man, with a finely shaped head and very forcible-looking blue eyes under rather shaggy eyebrows. His hair was fair, somewhat reddish on the whiskers and moustache. Later in life, when his hair was turning white, he wore a closely-clipped beard. It was a strong, determined-looking face, with those arresting eyes, making him a noticeable and dominant figure in any assembly. But the eyes could dance with fun on occasions, and his good Scot's tongue was kindly as well as outspoken. He remained sturdy and energetic to the last, although he was seventy-three years of age a few days before the motor accident in which he was instantly killed on March 16, 1914.

John Murray was a man of upright character and downright speech. He was apt to tell you what he thought of you, or anyone else, in plain and emphatic language without fear or favour. Some people of more conventional habits may have been shocked or offended at times; but the better one knew him the more one came to appreciate and admire his transparent honesty of thought and speech, his most uncommon "common sense," his purity of motive and directness of purpose, and his genuine kindness and good-heartedness, especially to all the young scientific men who worked with or under him, and whom he in large measure trained. He was absolutely free of all guile and humbug of any kind, and he had no sympathy with intrigue or vacillation.

I may appropriately conclude this short account of John Murray's life and work with a few sentences quoted from an appreciation (*Nature*, 1914, p. 89) by his old friend, and former teacher, Sir Archibald Geikie:

"Sir John Murray's devotion to science and his sagacity in following out the branches of inquiry which he resolved to pursue, were not more conspicuous than his warm sympathy with every line of investigation that seemed to promise further discoveries. He was an eminently broad-minded naturalist to whom the whole wide domain of nature was of interest. Full of originality and suggestiveness, he not only struck out into new paths for himself, but pointed them out to others, especially to younger men, whom he encouraged and assisted. His genial nature, his sense of humour, his generous helpfulness, and a certain delightful boyishness which he retained to the last, endeared him to a wide circle of friends, who will long miss his kindly and cheery presence."

10. THE LABORATORY COMES TO THE SEA

At the time that the *Challenger* put to sea, biologists were just beginning to realize that the greatest revolution that their science had ever known was in the making. In 1859 Charles Darwin had published *The Origin of Species*. Now it shook the very foundations of the science of life.

Its influence on man's view of the sea creatures was especially strong, for even then it was suspected that life had begun in the oceans. Thus a flood of new questions arose, and it became clear that many of them could best be answered through the study of the simple animals of the sea. What was the true origin of these ocean forms? What was the relation of one group to another? What of the structure and function of living forms in the light of the long process of evolutionary change? What of the economy of the important commercial fishes, and of oysters, lobsters, and crabs?

Questions like these flooded the mind of the young German zoologist Anton Dohrn in the decade that followed Darwin's publication. Dohrn had studied marine biology at Messina in

Sicily in the 1860's, and when he put Darwin's questions against the answers that life in the Mediterranean might yield, he conceived a magnificent idea: to establish an international marine research station somewhere on Italian shores. Ultimately he chose Naples. There, in 1872 he made arrangements to occupy a building in the Villa Nazionale, overlooking the sparkling bay; and the years that followed amply proved the wisdom of his choice. He had to overcome many difficulties—especially financial—but he fought his way to creating the premier institution of its kind in the world. Within a few decades he could point to more than fifteen hundred naturalists who had come to work in his laboratories and returned to share their new knowledge with students and colleagues throughout the world.

One of these was Charles A. Kofoid, who came from the Berkeley campus of the University of California, and who wrote a book for the United States Department of Education called *The Biological Stations of Europe*. Kofoid opened his book with this paragraph:

"A decade after the publication of *The Origin of Species* the fructifying influence of the new idea had not only brought new zest to classroom instruction and made the biological laboratory an inseparable part of the equipment of a university, but it also sent the investigator forth from the museum and laboratory to that greatest of the arenas of evolution, the seashore. Imbued with the idea that the great problems in biology which the theory of natural selection had brought to light, could be solved more speedily and more satisfactorily at the seashore than elsewhere, a young enthusiast, fresh from the laboratories of Jena, set out to establish and equip a great marine observatory on the shores of the Mediterranean. Others had preceded young Dohrn, but none so clearly acknowledged his debt to Darwin. A host followed him, not only to Naples, but to scores of stations springing up in other lands and on far-distant shores."

Thus Anton Dohrn "brought the laboratory to the seashore and the sea within the walls of the laboratory." Today the seaside laboratories, or marine stations as they are usually called, number in the hundreds, but Naples remains a pioneer.

Anton Dohrn was born in Stettin, Germany (now Szczecin, Poland), and studied at the University of Jena under Ernst Haeckel, the man who persuaded the German zoologists to Darwin's views; and quite naturally, Dohrn did his research on the development (embryology) and structure (morphology) of a number of animal groups. This led him to later work on the origin of vertebrate species. In *Founders of Oceanography and Their Work,* Sir William A. Herdman tells us that "in addition to being a man of ideas and a great organizer and administrator [Dohrn] was an eminent zoologist, and produced a large amount of first rate original research. The great work of his life was to prove that vertebrates were derived from Chaetopod worms [segmented marine worms that have heads with eyes, and definite blood systems], and that their characteristic features were not newly acquired, but were modifications of other organs which had in the ancestral worms some different function to perform."

Such a contribution to evolutionary theory would have been enough to make the reputation of any man, but it is not primarily for this that Anton Dohrn is best known—nor to which he gave his greatest effort. That was in his founding of the great Stazione Zoologica, to which he contributed most of his personal fortune; and in his wise direction of its growth and influence over a period of thirty-five years. He was indeed "its founder, benefactor, director, the source of its activities, and the source of its inspiration."

When the first building became available for use in 1874, Dohrn's problems in financing it were by no means at an end, but he had planned well, and as time passed, support began to flow in from many quarters. Some funds came from outright grants by various European governments, but most came from the system of "tables" that Dohrn had devised.

A "table" was not a laboratory table in the usual sense, but rather a small laboratory room, or a section of a larger room, which was assigned to a visiting researcher for a period of one year for a fee of £100 (about $500 in those days). Before long, visitors from all over Europe were at work at some fifty tables, supported by their own governments or, more often, by their universities. Thus at one time Germany supported eleven tables, Italy supported thirteen, Russia four, Austria two, Belgium two, Holland two, and Hungary, Switzerland, and the Roumanian Academy of Sciences one each. English tables were supported by Oxford and Cambridge universities and the British Association for the Advancement of Science; and American tables were supported at various times by the Smithsonian Institution, the Carnegie Foundation, Columbia University, and the University of Pennsylvania. Later, "The Naples Table Association for Promoting Laboratory Research by Women" was founded by Dr. Ida Hyde.

These international occupants of Naples tables were provided with every facility. "Upon the morning of arrival," Professor Kofoid reported, "the expected naturalist finds upon his table a collection of preserved material, it may be, or a dish of fresh sea water with some brilliantly colored squirming inhabitant of the subtropical Gulf of Naples, brought that morning by a barefooted fisherman from Margellina, Prosilippo, Procida, or perhaps distant Capri or Ischia. But one may rest assured that the material will be there unless Neptune or Boreas are angry, or the 'sirocco' drives even the amphibious Neapolitan fisherman to land. Every morning at 10 o'clock the ten or more fishermen of the station, with buckets or baskets poised gracefully on their heads, march into the court and receiving room of the station with their prizes from the sea—scarlet starfishes, orange feather stars, red and black sea-cucumbers, sea-urchins, squirming serpent stars, and bristling purple sea-urchins; or it may be a bit of red coral or a wriggling creeping octopus. No less interesting and wonderful for beauty of form and color are the translucent, shimmering colors of the violet Vellela, the band-like phosphorescent Venus's girdle, or the brilliant domes of the medusae. Equally prized by the naturalist is the great array of less highly colored and seemingly uninteresting worms, crabs, and amorphous sponges, or even the brown slime that the nets of finest silk sift from the blue waters of the bay."

To work with these living materials, the researchers were provided with every laboratory convenience: reagents, preservatives, glassware, aquaria, and drawing materials, including the camera lucida. Costly chemicals were also available, such as gold and platinum chlorides, osmic acid, cocaine, and so on, although workers were charged for these if they used them in large amounts.

The second source of income that Dohrn arranged was the admission fees to his magnificent aquarium. Attracted by one of the most glorious settings in the world, tourists came in numbers, and the Stazione Zoologica was a principal center of interest. The Villa Nazionale, in which the station is located, is a beautiful public garden on the shore of the Bay of Naples. Within the garden, palm leaves rattle softly in the tropic breeze, succulent cacti soften the brighter hues of green, and aloes unfold their red and yellow flowers. Fountains send streams of water to sparkle in the sunlight, and small temples of white marble look out over a panorama that stretches from Vesuvius past Sorrento and Capri to Ischia. Within the Italian Renaissance building of tufa masonry, the visitor sees aquaria of one, two, or even three stories, with marble floors and heavy plate-glass sides. In their native sea water swim, ooze, or sit marine creatures of every shape and color—a profusion to keep a visitor fascinated all day.

Dohrn also derived income from the specimens that he sent in beautifully preserved condition to universities and research centers throughout the world.

But the dominating focus of activity at Naples has always been upon research, especially in marine zoology in its broadest sense; and many original experimental studies have been carried out by noted authorities on systematics, morphology, and embryology. The results of this work have been published in richly illustrated monographs on all aspects of the Mediterranean fauna and flora. Visiting investigators have also been free to publish their findings elsewhere, and accounts of work done at Naples have appeared in many languages throughout the world's scientific journals.

Anton Dohrn opened his second building in 1890, and a third in 1907. The last was given over entirely to physiology, for by then many physiologists saw that problems concerning the physiology of animals generally could best be approached through the study of simple animals of the sea.

Dohrn died in Munich in 1909, but not before he had earned the respect and affection of biologists throughout the world. And the influence of his work lived on. As late as 1956 Dr. C. M. Yonge, Regius Professor of Zoology in the University of Glasgow, wrote in *Science Progress,* "The laboratory at Naples is unique in virtue of the character and scientific standing of its founder. By the institution of 'tables' financed by different countries, Anton Dohrn achieved his ambition of making the Stazione Zoologica an international institution. It quickly became not only a great center for original research in marine biology but also the meeting ground for zoologists, for general and comparative physiologists, and for others interested in the various aspects of the science of the sea. . . . For forty years, until the outbreak of the first world war, the Stazione Zoologica at Naples remained, beyond question, the premier marine station in the world."

In America, the first marine station was opened in 1873, within a year of the time that Anton Dohrn's first building was occupied in Naples. On Penikese Island in Buzzards Bay, Louis Agassiz, as we have seen, opened the Anderson School of Natural History, and the tradition it set was to live on also. Though Agassiz died that same fall, within a dozen years the founders of the Marine Biological Laboratory at Woods Hole—the direct successor of Penikese—had met "to perfect plans for the organization of a permanent seaside laboratory, to elect trustees, and to devise ways and means for collecting the necessary funds."

The needs for such a laboratory that the founders saw were two: the need to make better use of ocean resources and the need to provide better facilities for fundamental research in biology than the university laboratories afforded.

As to the economic need, it is well worth emphasizing that many years ago men were realizing that their industrial activities could be a threat to the environment. The so-called ecology movement of the 1970's is nothing new. In *The Biological Stations of Europe,* Charles Kofoid wrote, "The utilization of machinery in fishing operations, the great improvements in transportation and marketing, and the increased demands for products of the sea for food have made imperative a scientific basis for aquiculture if the harvests of the sea are to be fully reaped and its resources to be maintained unimpaired for the future. The pollution of the lakes and streams by municipal, industrial, and chemical wastes, and the ill effects of overfishing threaten to exterminate the native fishes and to destroy entirely a source of great pleasure and profit

The Marine Biological Laboratory at Woods Hole, Massachusetts. The main laboratory building is directly behind the two short piers.

to the human race. To meet these exigencies created by the rapid advance of our civilization a new type of station is now in process of evolution, one, moreover, which is no longer merely a biological station, but rather a station equipped for the solution of biological problems with the aid of all pertinent sciences."

This concept of a new kind of marine biological research station was fostered, developed, and brought into bloom by one man—Dr. Charles Otis Whitman—who at one time had occupied a table at Naples and who felt a deep friendship and admiration for Anton Dohrn. But Whitman was also deeply influenced by Louis Agassiz, and when he was named director of the Marine Biological Laboratory at Woods Hole in 1888, he visualized an institution quite different from Dohrn's Stazione Zoologica. He thought that Woods Hole should be a place where every investigator would have a free choice of the project on which he would work, in whatever field of biology it might happen to be, entirely uninfluenced by direction of any sort from the administrative staff. The result of this policy was that more and more faculty members from universities throughout the United States (and later from overseas as well) came to Woods Hole to continue their ongoing work during the summer months.

This work might concern an animal that had nothing to do with the sea, such as Adele M. Fielde's study of the behavior of ants in relation to their sense-discrimination ability, and her theory that each joint of an ant's antenna has a special function in distinguishing odors—in short, that an ant's antenna is a series of specialized noses! Or it might concern the multiple embryos that sea-urchin eggs produced when Jacques Loeb put them into diluted sea water. It did not matter to Whitman.

When a new investigator first came to Woods Hole, Whitman would ask, "What is your beast?" And the only restriction he would put on the visitor's work would be, "Nobody should meddle with my beast without my permission, and certainly not in the case of my colleagues!"

Typical of the breadth of subject matter that Whitman encouraged was his own beast—the pigeon, in which he studied the heredity of color patterns. He surveyed color patterns in six hundred wild species and two hundred domestic races; and in his own yard he raised and hybridized some forty wild species. Seeing that he tended these birds himself every day, Frank R. Lillie never understood how Whitman remained so impeccably dressed.

Charles Whitman came to Woods Hole with outstanding

The group of researchers that came to work at the Woods Hole Marine Biological Laboratory in the summer of 1895.

qualifications to serve as its first director. He had attended Louis Agassiz's first session at Penikese Island in 1873 and returned for further work there in marine biology in 1874. He took his Ph.D. in zoology in 1878 under Leuckart at the University of Leipzig, and immediately accepted a two-year appointment as professor of zoology at the University of Tokyo. On his way back to the United States he was a special guest of Anton Dohrn's at Naples. Back at home, he served in the Museum of Natural History at Harvard, was director of the Lake Laboratory in Milwaukee, Wisconsin, and finally found his niche as professor of zoology at Clark University in Worcester, Massachusetts. He combined that professorship with his directorship at Woods Hole, an arrangement that has since become traditional.

When Whitman first accepted the directorship in 1888, neither he nor the trustees had more than the most general ideas about how the laboratory should function. But Whitman maintained a continuing correspondence with both American and European biologists, and by 1898 he was able to crystallize the ideas that were to represent the cornerstone of MBL philosophy in the years to come. Writing on "Some of the Functions and Features of a Biological Station," he said:

> It now remains to briefly sketch the general character and to emphasize some of the leading features to be represented by a biological station.
>
> The first requisite is capacity for growth in all directions consistent with the symmetrical development of biology as a whole. The second requisite is the union of the two functions, research and instruction, in such relations as will best hold the work and the workers in the natural co-ordination essential in scientific progress and to individual development. It is on this basis that I would construct the ideal and test every practical issue.
>
> A scheme that excluded all limitations except such as nature prescribes is just broad enough to take in the science, and that does not strike me as at all extravagant or even as exceeding by a hair's breadth the essentials. Whoever feels it an advantage to be fettered by self-imposed limitations will part company with us here. If any one is troubled with the question: Of what use is an ideal too large to be realized? I will answer at once. It is the merit of this ideal that it can be realized just as every sound ideal can be realized, only by gradual growth. An ideal that could be realized all at once would exclude growth and leave nothing to be done but to work in grooves. This is precisely the danger we are seeking to avoid.
>
> The two fundamental requisites which I have just defined scarcely need any amplification. Their implications, however, are far-reaching, and I may, therefore, point out a little more explicitly what is involved. I have made use of the term "biological station" in preference to those in more common use, for the reason that my ideal rejects every artificial limitation that might check growth or force a one-sided development. I have in mind, then, not a station devoted exclusively to zoology, or exclusively to botany, or exclusively to physiology; not a station limited to the study of marine plants and animals; not a lacustral station dealing with land and fresh-water faunas and floras; not a station limited to experimental work, but a genuine biological station, embrac-

ing all its important divisions, absolutely free of every artificial restriction.[1]

Now that is a scheme that can grow just as fast as biology grows, and I am of the opinion that nothing short of it could ever adequately represent a national center of instruction and research in biology. Vast as the scheme is, at least in its possibilities, it is a true germ, all the principal parts of which could be realized in respectable beginnings in a very few years and at no enormous expense. With scarcely anything beyond our hands to work with, we have already succeeded in getting zoology and botany well started at Woods Hole, and physiology is ready to follow.

How fully these aims were met over the years is well shown by the many MBL publications themselves: *The Annual Reports,* which also contain reports of evening lectures and shorter scientific papers; and the *Biological Bulletin,* which has published the results of Woods Hole research since 1902.

That, then, is the formal side of Woods Hole. It also has its invaluable informal side. As Frank R. Lillie put it, "All science is a social function, in that it is dependent upon cooperation. Every advance in knowledge is built upon that which has gone before. Every worker is indebted to other workers. Comparison, confirmation, or criticism of the work of others enters into all scientific progress. . . . At Woods Hole . . . the spirit of frank and open discussion has grown with the intimate association of the workers; and the influence of this Laboratory in promoting peaceful cooperation in biology is one of its major by-products."

As early as 1902, Whitman began to withdraw from the affairs of the Laboratory, although he retained his title as director until 1908. He made a visit of a few days in 1909, and finally in 1910 he left, according to some lines he had once quoted, for

*"That undiscovered country from whose bourne
No traveler returns."*

The world's third major marine laboratory was established in La Jolla, California. It had the most precarious start of all, but it was destined to become the largest. Like Naples and Woods Hole, its greatest debt—although it certainly has many—is also

---

[1] A visitor from a European university, accustomed to a very different attitude, once said in astonishment, "Well, this must be a heaven of freedom!"

to the courage and determination of one man. That man was William Emerson Ritter.

Ritter was born in Wisconsin, became a teacher there, and as his active mind sought out new knowledge of "the science of life in its broadest sense," he read a textbook on geology by Joseph Le Conte, then professor of natural history at the quite insignificant University of California in Berkeley. Affectionately known to the Berkeley of those days as "Little Joe Le Conte," he was a master teacher, and his enthusiasm for his subject imbued even a geology text with vibrant life. Ritter was enthralled, and as soon as he could break free from his teaching, he left Wisconsin for the shores of San Francisco Bay.

The trip cost him most of his slim savings, and he taught for a year in the sun-drenched grape and raisin center of Fresno. Returning to Berkeley in the summer of 1886 (Berkeley's cool summer fogs provided ideal study conditions in August), William Ritter enrolled as a major in zoology and earned his B.S. degree two years later. The next year he did graduate work at Berkeley, and then was awarded a scholarship at Harvard. For his Ph.D. research he chose to work on the rudimentary eyes of the flesh-colored fish known as the blind goby, which spends its life in dark crevices in rocky California coastal areas. (How Louis Agassiz would have delighted in this study of *Typhlogobius californiensis!*)

Just as Ritter had earned his Ph.D. from Harvard, the University of California decided to form a new department of zoology, and Ritter was offered the chairmanship. He took it eagerly, and from 1891 until his retirement as Director of the Scripps Institution of Oceanography in 1923, he was a member of the faculty of the University of California.

The germinal idea of a University of California seaside laboratory took shape with young Professor Ritter's ambition to make a thorough, systematic, and continuing biological survey of the Pacific Ocean near the coast of California. He wanted to extend that study to the deeper waters beyond the intertidal zone, and he also realized that the chemical and physical properties of the ocean waters must influence the kinds of animal communities to be found in various locations. Thus from the beginning, his ideas differed from those at Naples and Woods Hole. He was aiming—even though somewhat vaguely at first—at a most broadly based institution of oceanography.

His dream was to be amply realized in the years to come, but he could hardly have had much faith in that when he took his

The Scripps Institution of Oceanography is behind and mostly to the right of the long pier. The main campus of the University of California at San Diego appears in the middle distance.

first group of seminar students to work in a tent near Pacific Grove, on the Monterey peninsula.

That was in the summer of 1892, and on an appropriation of $200 granted by the Zoology Department, Ritter and his handful of students collected, sorted, and classified the sea animals that they found on the littoral of Monterey Bay. The following summer, they pitched their tent at Avalon on Santa Catalina Island, and in later summers they worked off San Pedro Harbor near Los Angeles, at San Diego, and during one memorable summer aboard E. H. Harriman's yacht, while exploring the waters of Puget Sound and much of the coastline northward toward Alaska.

Finally, in 1905, a permanent building for the summer work was decided on, and "The Little Green Laboratory at the Cove" was built at La Jolla at a cost of $992. This was an impressive sum in those days, and it was raised to a large extent through the effort of E. W. Scripps, the publisher, and his gracious sister,

Ellen Scripps. Business and professional people throughout the San Diego area also contributed, as members of the San Diego Marine Biological Association. However, as time passed it became clear to even the most generous sponsors that the management of an enterprise as ambitious as they hoped for would be beyond their scope, and in 1912 the affairs and property of the Association were transferred to the University of California. The official designation at that time was "The Scripps Institution for Biological Research of the University of California."

William Ritter continued as director until his retirement in 1923. Five years later, Dr. Fred Baker, the venerated physician of San Diego who had helped the laboratory through all its early struggles, dedicated a portrait of Ritter that had been commissioned by Ellen Scripps. Speaking of "Dr. Ritter and the Founding of the Scripps Institution of Oceanography," he said:

"I believe . . . that few institutions of this character have

FLIP, a 355-foot Floating Instrument Platform, was developed by the Marine Physical Laboratory at Scripps Institution of Oceanography, University of California, La Jolla. She is photographed here while "flipping" from the horizontal to the vertical position, where she will act as an extremely stable platform from which scientists will conduct oceanographic studies. FLIP has no motive power of her own and must be towed to a research site in the horizontal position. Once on station, her ballast tanks are flooded and she "flips" vertically, keeping her platform 55 feet above water.

The University of Washington Oceanographic Laboratory, Friday Harbor, San Juan Island.

grown up which more fully express the ideas and ideals of one man than does this one. Rarely assertive, always happily good-natured, ready at all times to listen to others and to defer to their judgment, nevertheless he was the dominating force which drove us all to the goal which he had set."

These, then, in brief outline, are the stories of the men who founded three of the greatest seaside laboratories of the world. Even to summarize the contributions that these laboratories made in the years that followed would take volumes, but in some of our later chapters we shall see a few of the more spectacular adventures that facilities of this kind have made possible.

Today seaside laboratories number in the hundreds; they are found throughout the world, and their missions vary from the most esoteric research in fundamental biology to practical ways of helping sports fishermen improve their catch.

One sample will serve to illustrate. Looking at the maps of the Pacific coast in *Between Pacific Tides,* first written in 1939 by Ed Ricketts and Jack Calvin, and now ably and sympathetically revised by Joel W. Hedgpeth, we see that marine stations run more or less continuously from the Nanaimo Biological Station on Vancouver Island to Escuela Superior de Ciencias Marinas in Ensenada, Baja California. Along that coastline we can count at least twenty stations, each with its own individual area of interest.

Can we see what it was that enticed men to set up all these laboratories by the sea? Scientific curiosity, we may suggest. But perhaps we can find something more. Philip Henry Gosse touched on it in 1853 in *A Naturalist's Rambles on the Devonshire Coast* when he wrote:

"The sea-side is never dull: other places soon tire us; we cannot always be admiring scenery, though ever so beautiful, and nobody stands gazing into a field, or on a hedgerow bank, though studded with the most lovely flowers, by the half-hour together. But we can and do stand watching the sea, and feel reluctant to leave it; the changes of the tide, and the ever-rolling, breaking and retiring waves are so much like the phenomena of life, that we look on with an interest and expectation akin to that with which we watch the proceedings of living beings."

11. ALEXANDER AGASSIZ: COPPER MINES AND CORAL REEFS

One day in the early spring of 1867, Alexander Agassiz met Charles W. Eliot on a Boston street. Eliot, then an obscure young professor of chemistry at the Massachusetts Institute of Technology, was soon to be named Harvard's youngest president. Agassiz, son of the world-famous Harvard naturalist Louis Agassiz, was serving as "agent" for the newly established Museum of Comparative Zoology at a salary of $1500 a year.

"Eliot," said young Agassiz, "I am going to Michigan for some years as superintendent of the Calumet and Hecla mines. I want to make some money; it is impossible to be a productive naturalist in this country without money. I am going to get some money if I can, and then I will be a naturalist. If I can succeed, I can then get my own papers and drawings printed and help my father at the Museum."

He did succeed, although the obstacles that he had to overcome were formidable. "Seldom, indeed," as his son later said in his biography, "have the aspirations of youth proved in such harmony with the achievements of maturity."

Alexander Agassiz was born in 1835 in Neuchâtel, Switzerland, the first of three children of Louis and Cécile Braun Agassiz. At that time his father was a struggling professor of natural history at Neuchâtel, and the family lived in a tiny apartment "in the most straitened circumstances." Young Alexander found much pleasure in long walking trips through the woods and fields around Lake Neuchâtel, and was a great collector, even at an early age. He soon learned how to preserve his specimens in alcohol, and he kept them carefully in a wardrobe drawer. One of his cousins recalls that whenever a playmate came too near these treasures he would cry, "Please don't touch my anatomy!"

In the spring of 1848, after his father had gone to America, and he had lived for nearly a year with his uncle in Freiburg, Germany, his father sent for him. Overjoyed at the prospect of going to the New World, he thought of the endless hours of violin practice that had been forced on him, especially of those chill wintry mornings in Freiburg Cathedral, when his teacher would rap his numbed knuckles after each mistake. He contrasted that with a land where he could do as he pleased, and to celebrate his coming departure he threw his violin on the floor and jumped on it.

Alexander Agassiz found the wild creatures of New England quite as fascinating as those in Switzerland and Germany. He was also introduced to the sea, for which he was to have a lifelong passion—although his first meeting with it was a disaster.

His father took him for a voyage on the Coast Survey vessel *Bibb,* and during a brisk seaway, Alexander fell down an open hatch. When they laid him out in the saloon, they thought he was dead, but he had broken no bones, and he lived to sail a hundred thousand miles over the oceans of the world.

Ashore during those early years, he spent much time in a ramshackle wooden shed in Cambridge, where his father kept his growing collection of animal specimens. That was the rude beginning of what was to become one of the greatest collections in the world: the Museum of Comparative Zoology at Harvard. In later years, Alexander was to give a fortune to its buildings, its specimens, and its publications.

Much of the Museum's collection was built up by exchanging local specimens with museums in other parts of the world. Agassiz forwarded one shipment, destined for the famed *Jardin des Plantes* in Paris, with the following letter of instructions:

GENTLEMEN:—

I forward today to you per Adams Express 9 boxes, 8 of which contain live-stock to be forwarded to Prof. H. Milne Edwards, Jardin des Plantes, Paris, France, care of H. L. Müller and Cie., Havre. The food of the animals is marked on the boxes, but to make sure I repeat here:

No. 1 contains Reptiles and needs no care except air.
    2 contains a Marten; needs scraps of meat and water.
    3 a Lynx; needs same food as No. 2.
    4 & 5 contain squirrels; need nuts, scraps of bread, corn and *water*.
    6 Woodchuck; eats turnips, raw potatoes, scraps of vegetables and *water*.
    7 Owl; needs only meat.
    8 Eagle; needs meat and water.

If these animals are fed once a day it is enough except the woodchuck, No. 6, which had better be fed twice. I send in the ninth box nuts, turnips and corn for the food of the squirrels and part of the food of the woodchuck. I suppose of course that scraps of bread and the necessary remnants of *fresh meat* can be obtained on board the steamer by the person who has charge of them. The lynx ought to be bothered as little as possible; it is a female with young and she is rather cross on that account. I suppose the cages of the lynx and of the woodchuck ought to be cleaned about once in four days, if they get offensive, and clean hay or straw put in. This can easily be done by means of a poker to scrape out the old hay.

Hoping that this small menagerie will have a favorable passage, I remain,

Yours very truly,
ALEX AGASSIZ.

The careful planning and attention to detail that we see here was an outstanding characteristic of Agassiz throughout his life.

One might guess that a man who was to lay the foundation for a fortune in copper before he was forty and who was later to become a world-famous naturalist, would have had an unusual college education. Alexander Agassiz surely did.

He entered Harvard in 1851, and graduated four years later with no particular honors, for he had slight interest in philosophy or the classics—subjects that were the core of the undergraduate curricula of those days. In mathematics and chemistry he did extremely well, and his marks in those subjects raised his average enough so that he could graduate twenty-fourth in a class of eighty-two.

Although he was only of medium height and rather slight, Agassiz pulled bow in the Harvard crew and formed a lifelong attachment to rowing. Indeed, for years after he had established his private laboratory in Newport, Rhode Island, his steam launch used to disappear each June—often along with Agassiz. It was at first suspected, and later confirmed, that they were then in New London, helping the Harvard crew prepare for the race with Yale.

In the fall of 1855 Agassiz entered the Lawrence Scientific School. Relieved now of the "subtle analysis of language and the dryness of grammatical hair-splitting," he threw his abundant energy into his engineering studies, and graduated *summa cum laude* in 1857. Thus prepared for a professional career offering prospects of a reasonable livelihood, he joined the Coast Survey. But some months of surveying activity on the Pacific Coast, and an independent venture in Panama, convinced him that his course did not lie in that direction.

He returned to Harvard in 1860, and there, under the influence of the exciting work that his father was doing at the Museum of Comparative Zoology, his interest in animals returned with renewed intensity. Back he went to the Lawrence Scientific School, where he took another advanced degree, this time in natural history. That was in 1862, and for the next five years he devoted himself to his work in the Museum, where he served both as scientist and as practical administrator, for his famous father was too deeply immersed in large affairs to bother with such details.

Soon after his return to Harvard from Panama, and his appointment as "agent" for the new Museum of Comparative Zoology, Agassiz married Anna Russell, the daughter of a prominent Boston family. He could not support his bride as he had hoped, and the meager Museum salary forced the young couple to live with the senior Agassizes on Quincy Street. But being poor was nothing new to Alexander, and that fall of 1860 began the happiest period of his life.

His marriage was a joyous one, the household of the senior Agassiz's was congenial and stimulating, and beyond his work at the Museum, Alexander found time for an impressive body of original work on the natural history of starfishes and urchins. His son later said in the *Letters and Recollections* that during this period he

> laid the foundation for all his purely zoological investigations.
>
> His summers, broken only by occasional scientific excursions to other portions of the New England Coast, were devoted to research at Nahant. . . . [There, the cottage] on a cliff overlooking the sea, with a rustic porch supported by unstripped fir logs, was in the last degree picturesque. In the rambling shed below the house, Agassiz and his father fitted tables, shelves, and the necessary glassware for a marine laboratory; and here they set up their microscopes. In a bight nearby, Alexander kept a dory which he used for collecting specimens. One of his step-cousins, then a small boy, used in after years to tell how he crept out to watch Agassiz skillfully launch his heavy boat through the surf, and admire the ease with which, without apparent effort, he ran it up the steep beach.
>
> The material for his studies was furnished by such animals as he could pick up among the rocks at low tide or catch with a scoop from his dory, varied by an occasional find brought in by some friendly fisherman.
>
> Among his best-known publications of this period was a series of papers on our common sea-urchins and starfishes (Echinoderms) which are found everywhere along our New England Coast hidden in crevices of the rocks or among the seaweed. Until Johannes Müller discovered where to find the young of these animals, who would have thought of looking for them among the minute transparent and phosphorescent organisms that float near the surface of the ocean, and transform the wake of one's boat on a still, dark summer night into a path of fire, and turn each tip of the oar blades into a swirl of molten gold? . . . It remained for Agassiz to make and complete a study of the embryology of these animals, partly by artificial fertilization and partly by collecting the tiny animals with a dip net.
>
> Agassiz, among other things, showed at this period that Cape Cod was the dividing line for many species of the marine animals frequenting the coast of the North Atlantic, and he also did much work on Echini [starfishes and sea urchins], in

preparation for his "Revision", published several years later. Most of the work of these earlier years was so fundamental in character that it is freely quoted, consciously or unconsciously, in all modern text-books of zoology.[1]

One might think that Agassiz would have been content to live out his life in these congenial surroundings, doing the kind of work that he most loved. Many gifted scientists have done just that. But Agassiz's brief venture into engineering would not let him rest, and he could not reconcile himself to the income that a purely scientific career could provide him.

He had already done a fine body of original work on New England seashore animals, and he had a portfolio of "the most delicate drawings from his own pencil," but the cost of publishing them was far beyond his means. Moreover, the financial problems of the struggling young Museum of Comparative Zoology constantly weighed on his mind, and he had no hope that his father's cavalier attitude toward such matters would ever change. Especially, he wanted a home of his own, more in keeping with the surroundings that his wife had known as the daughter of a prosperous Boston importer. But just as he seemed to be caught in a fruitless situation, there came a new turn of events.

His brother-in-law, Quincy A. Shaw, hard-pressed to reverse a deteriorating situation at the Calumet and Hecla copper mines in northern Michigan, offered Agassiz the managership of the mines. Cautious, Agassiz visited the mines on several tours of inspection, and his engineering training told him that the mines were in a bad way indeed. But he also saw that they were potentially very rich, and he not only accepted the managership but also borrowed heavily to buy shares in the venture.

The story of his next two years in the harsh Keweenaw copper country of northern Michigan is one of a continuing struggle against the follies and the sharp dealings of other men, against the refractory ore of the mines, against the near impossibility of operating in the deep snows of winter, against roller mills that failed to produce, and against a misfit locomotive.

From his headquarters at the mine property, Agassiz kept Shaw fully informed of his successes and his failures as they occurred by turns, and in November of 1867 he wrote:

"I have just received an awful rap over the knuckles. The

---

[1] That is, textbooks that were modern in 1913, when *Letters and Recollections* was published.

locomotive and track are not of one gauge; locomotive is *one inch too narrow*. This is perfectly infernal. First comes snow; and now, just as we had managed by dint of sweeping to lay a temporary track around the trestle of head of incline, to get locomotive up and push road to mine, we are brought to a dead standstill by this mistake and shall have to relay one rail whole distance from Lake and alter the axles of all our cars (which fit admirably) to the new gauge. This is no fool's job and will, I am afraid, delay us greatly."

But while difficulty after difficulty arose, Agassiz met each as it came, not always calmly, for his temper could flare up sharply, but with steady determination and unusual common sense.

"We are accustomed to think," his son says, "of the man who devotes himself to pure science as aloof from the world, with but little interest or ability in the practical concerns of our complex modern civilization. Agassiz is a striking exception to whatever truth there may be in such a belief, for while engrossed in his scientific life, he at the same time produced a wonderful practical achievement.

"Called as a last resort to prop up a falling enterprise, he transformed it into one of the most prosperous and extensive mines known in the history of industry."

Agassiz achieved this success in an amazingly brief two years, and at the end of that time he could leave the day-to-day operations to competent men whom he had trained. He returned to Harvard and to his marine studies with a sigh of relief—this time with the knowledge that he would at last have the funds to pursue his scientific studies in whatever direction his interests might lead him, and to publish his results in appropriate form.

He was elected president of Calumet and Hecla, and in that position he visited the mines every spring and fall and he also kept in close touch with the corporate offices in Boston. Otherwise he was free to roam the oceans of the world and to bring back new observations about life in the seas for summer study and evaluation in his private laboratory at Castle Hill in Newport.

Upon his return to Cambridge late in 1868, Agassiz resumed his correspondence with prominent naturalists throughout the world, among them Charles Darwin, Ernst Haeckel, Sir John Murray, Sir Wyville Thomson, F. A. Forel, Fritz Müller, T. H. Huxley, and A. Milne Edwards. At the close of the Franco-Prussian War, he wrote to Professor Edwards at the Jardin des Plantes in Paris:

"Everything that you will kindly tell me about the conditions of the Jardin will be most interesting, for no one here can form an idea of what you must have suffered during that abominable siege. Every one rejoices that peace is at length declared, and I am sure that no one can sympathize more than the Americans with all that you must have undergone during that horrible war, which seems to recall the dark ages rather than the nineteenth century....

"I trust that this will not lead to a military epoch in Europe, and that peace will at length teach the natives of Europe that all progress is impossible without the advancement and cultivation of science."

Soon after his return to Cambridge, Agassiz also began work

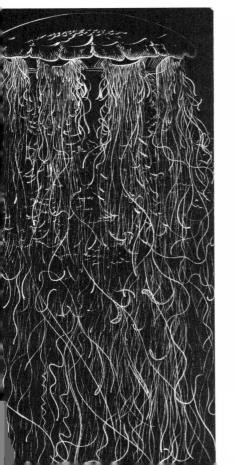

A giant jellyfish, Cyanea arctica, *painstakingly drawn by Alexander Agassiz. The illustration is greatly reduced from the natural size of the animal.*

Profile of a rocky beach in Massachusetts, as sketched by Alexander Agassiz. A beautiful study of the essence of marine ecology.

with his stepmother[2] on a delightful little seashore book that they published in 1871 under the title *Seaside Studies in Natural History; Marine Animals of Massachusetts Bay; Radiates*. Like Alexander, Elizabeth Cary Agassiz was fascinated by the animals of the seashore, and it was quite natural that she and Alexander should write the book together, she doing the words with the aid of his observations and notes, and he doing the intricate drawings of polyps, acalephs, and echinoderms. Written in popular style, the book was a frequent companion of non-specialist visitors to the shore; at the same time its careful attention to the principles of scientific classification and its detailed descriptions of the radiates, gave it a wide usefulness to more serious students and teachers.

Agassiz was thus on his way to a fortune, happily devoted to the work he most loved, and in close touch with the prominent naturalists of the day, when a double tragedy overtook him.

On December 14, 1873, his father died, and the founder and champion of the Museum of Comparative Zoology was gone. Eight days later, just before Christmas, his young wife Anna succumbed to pneumonia. To Huxley he wrote, "Few young men have reached my age [thirty-eight] and have attained, as it were, all their ambition might desire, and yet the one thing which I crave for and which I want to keep me interested in what is going on, is wanting. How gladly I would exchange all that I have for what I have lost. But I will not burden you with my sorrows."

Agassiz's son says that he was never again the same man, but that his "intellectual activities were undimmed, and he . . . pursued the secrets of nature in [a] steadfast endeavour to increase the sum of human knowledge." Agassiz himself voiced the principle that would guide him from then on: "To live our lives as they have been made for us, and live in hope, do the best we can, work hard, and have as many interests as we can in what is going on around us."

Agassiz pursued such interests over all the world's oceans. By turns he investigated ocean animals, ocean currents, and coral reefs, in the Caribbean Sea, the South Seas, the Eastern Tropical Pacific, and the seas around the Bahamas, Cuba, Bermuda, Florida,

---

[2] While the term is factually correct, "stepmother" sometimes has connotations that are far removed from the relationship that had developed between Elizabeth Cary Agassiz and Alexander. She married Alexander's father in 1850, while Alexander was still a boy in Europe, and she made the Agassiz home in Cambridge "one of the centres of the intellectual life of the day, and [going] straight to the heart of the motherless boy, she stayed there for the rest of her life, his devoted friend and companion." (G. R. Agassiz: *Letters and Recollections*.)

the Fiji Islands, the Maldives, and the Great Barrier Reef of Australia. In *Founders of Oceanography and Their Work*, Sir William Herdman says, "His voyages covered more than 100,000 miles in tropical seas, and it has been said that he personally has run more lines of investigation across the great oceans and has made more deep-sea soundings than all other oceanographers taken together."

Most of his investigations up to 1892 were focused on marine zoology, but in that year Agassiz became deeply interested in the way that coral reefs and coral islands are formed. It is true, of course, that the corals themselves are marine animals—tiny polyps that range from the size of a pinhead to that of a pea—but the problem of how coral reefs are formed also involves the broader question of the forces that have shaped the surface of our planet in recent geological times. It was to the investigation of coral reefs throughout the world that Agassiz was to devote the last two decades of his life.

More than fifty years earlier, men of science had been at a loss to explain the formation of coral reefs—the ring-shaped reefs that enclose shallow lagoons, sometimes with an island in the center of the lagoon and sometimes without. The puzzle lay in the fact that corals do not grow below a depth of about ten fathoms. Thus the reefs could not have emerged from the ocean deeps, but must have begun to grow upon some kind of supporting base no more than sixty feet below the surface.

After Charles Darwin returned from the voyage of the *Beagle*, he proposed the first general theory, and published it in 1842 in *The Structure and Distribution of Coral Reefs*. There he proposed the idea that atolls always begin with a thin layer of coral growing around the submerged shoulders of an island (say, at a depth of about sixty feet all around the shoreline). Then, as the coral continued to grow upward year after year, it would eventually form a ring around the island, and so we would see a coral reef surrounding the island, with a circular lagoon between the island and the reef. However, many such reefs had been seen in which the central islands were very low, and often, too, no island was to be seen at all; in that case the reef simply surrounded a shallow circular lagoon. To explain the disappearance of the islands, Darwin supposed that in every case the central island had subsided. At the beginning of the process, then, an island would be surrounded by a ring of coral far below the surface; as the coral grew and the island sank, an intermediate stage would occur in which the reef had reached the surface and only the tip of the island remained above water; in the final stage the island would have sunk beneath the waves and only the reef would remain, surrounding the shallow lagoon.

The islands that Darwin studied were of volcanic origin, and were thus in areas where the earth's crust was unusually unstable. Massive forces below had spewed up lava over the centuries until an island had finally appeared above the sea, and these volcanic cones had erupted further to build up an enormous weight of volcanic rock pressing down upon the still unstable ocean floor. What more logical than to assume that as the volcanic activity in the earth's interior diminished, the island would gradually subside?

It was on this basis that Darwin thought that wherever coral atolls would be found, the central island upon which the coral had first begun to grow would be gradually subsiding, and he went so far as to extend the reasoning to atolls and barrier reefs everywhere.

Darwin's reputation was such that it was many years before anyone questioned his seemingly logical theory. However, when Sir John Murray returned from the *Challenger* expedition in 1876, he went through his notes on all the coral atolls that he had seen and was struck by the fact that in no case was there any evidence of subsidence. If locations could be found where subsidence did occur, he thought, Darwin's theory might apply; but there were certainly many situations in which it did not. Reef areas were known in the West Indies where the earth's crust was actually rising rather than sinking.

All this was well known to Agassiz, and when he was in Edinburgh helping to distribute the collections of the *Challenger* among specialists[3] throughout the world, he and Murray had long talks about the problem. So as early as 1877 Agassiz decided to make a thorough investigation of the coral reef problem throughout the tropic seas.

---

[3] Agassiz himself wrote the report about the echini. When he had finished he wrote to Wyville Thomson, "I felt when I got through that I never wanted to see another sea urchin and hoped that they would gradually become extinct."

This X-radio graph of a specimen of Favia *coral from Einwetok atoll clearly shows the pattern of annual growth rings. The rate of growth can be seen from the fact that the reproduction is natural size.*

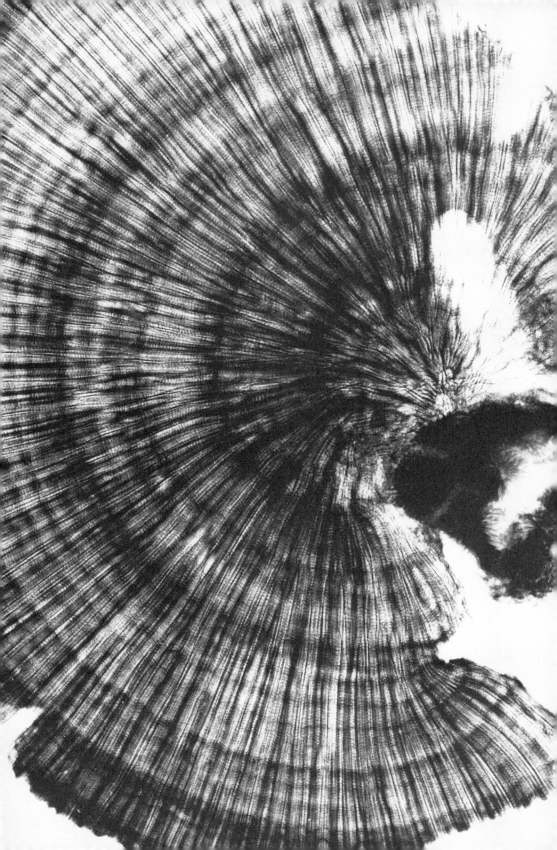

The expeditions that he carried out covered a period of ten years. In 1892 he was in the Bahamas and in Cuban waters; in 1894 in Bermuda and Florida; in 1895 in Hawaii; in 1896 at the Great Barrier Reef; in 1897 and 1898 in Fiji; and finally in 1902 in the Maldives, southwest of Ceylon.

In not one case did he find an atoll that conformed to Darwin's theory. After Agassiz had published his findings from the Hawaiian Islands, the adherents of Darwin's theory (most of whom had never seen a coral atoll) protested that Hawaii was simply an exceptional area. "This criticism," his son wrote, "he was destined to meet with such tantalizing frequency in after years, that his subsequent expeditions might almost be called a search for a typical coral region."

He never found one anywhere. And as a result of these extensive investigations he concluded that there was no single, simple explanation. In some cases, perhaps, Darwin's theory might well hold. In others, especially where it was clear that no subsidence of the land had occurred, some other explanation must be found. Banks and shoals might be built up by the accumulation of silt washing into the sea, by the deposit of shells of marine animals, by the erosion of volcanic islands, or even by wind-blown limestone drifting over the shallows. Once such shoals had been built up to a suitable depth, the corals could then take up their work upon the bases so prepared.

According to Sir William A. Herdman, Dr. A. G. Mayer, who had gone with Agassiz on several of his voyages and who was director of the Carnegie Institution Research Laboratory on the Tortugas, went so far as to say, "I believe science will come to see that he succeeded in showing that Darwin's simple theory of the formation of atolls does not hold in any part of the world."

Of course, disagreements between men who hold different theories have been characteristic of the history of science, and in many cases they led to bitter controversy and even personal vituperation, such as Darwin himself encountered when he advanced his theory of evolution. But there was no animosity between Darwin and Agassiz. Their correspondence shows that each was willing to give full consideration to the ideas of the other.

Writing to Darwin from the Tortugas, Agassiz said: "It is very natural you should be on my mind, as I am in the midst of corals. . . . The greater part of my time I spend in running round inside the reef in the launch and getting at the distribution of the different genera of corals. The number of species here is not great, so it makes their mapping out a simple matter. The Tortugas being the very last of the Florida reefs I find much that has not been noticed before and [that] helps to explain . . . the formation of the reefs. . . . In tracing the growth of the reefs and the formation of the Peninsula, I have come across no signs of any elevation. Everything, on the contrary, tends to show that the immense plateau which forms the base upon which the Peninsula of Florida is formed, was built up by the debris of animal remains, —Mollusks, Corals, Echinoderms, etc. (after it had originally reached a certain depth in the ocean), until it reached the proper height for corals to flourish. This here is not much deeper than seven to eight fathoms; generally six fathoms marks the limit."

To this Darwin replied: "From the few dredgings made in the Beagle in the South Temperate regions, I concluded that shells, the smaller corals, etc., etc., decayed and were dissolved, when not protected by depositions of sediment; and sediment could not accumulate in the open ocean. . . . Pray forgive me for troubling you at such length, but it has occurred to me that you might be disposed to give, after your wide experience your judgment. If I am wrong, the sooner I am knocked over the head and annihilated, so much the better. It still seems to me a marvelous thing that there should not have been much and long-continued subsidence in the beds of the great oceans."

In addition to Agassiz's characteristic open-mindedness, he was a sympathetic and generous man, and recipients of his generosity often attested to it.

In *Founders of Oceanography and Their Work,* Sir William A. Herdman recalled the meeting of their two research vessels in Ceylon. "It was during Agassiz's Maldive trip in the winter of 1901–2 that I had a most interesting interview with him. I had met him before in Edinburgh, had visited him in his Newport laboratory, and again since, at Harvard, but at Colombo in Ceylon in January, 1902, we spent a long day and evening together. He had just returned from his Maldive expedition and I was just starting mine to the pearl banks in the Gulf of Manaar. Our two steamers, both chartered from the British India Co., lay at anchor side by side in the harbour, and we dined on shore that evening and discussed coral reefs, tropical seas, and marine biology in general. My expedition profited greatly by that chance encounter, for the next morning, before I sailed Agassiz had shipped from his vessel to mine some 600 fathoms of steel dredging wire and an odd assortment of store bottles and tubes left over from his expedition.

"I had thought of him before as a quiet, reserved man of great determination and ability. It has been said of him in America: 'He was a colossal leader of great enterprise fully as much as he was a man of science.' But at that time in Colombo, and also since, I have felt that he was also very thoughtful for others and of a kindly and generous disposition."

Also, in California in the early 1900's, when "The Little Green Laboratory at the Cove," which was to become the Scripps Institution of Oceanography, was struggling for its very existence, Agassiz gave a helping hand. In *Scripps Institution of Oceanography*, Helen Raitt and Beatrice Moulton recall: "Perhaps the most notable contribution came from Alexander Agassiz, the eminent Harvard-based biologist. . . . Learning of the station from C. A. Kofoid, who was a member of Agassiz's six-month South Sea expedition of 1904–05, the noted scientist paid San Diego a visit in March of 1905. Agassiz talked with E. W. Scripps and other members of the association at Miramar Ranch, looked over the La Jolla site, and promised substantial support. He gave books and scientific apparatus worth $1500 to $2000, as well as an immeasurable boost to the morale of everyone concerned."

Agassiz's largest contribution by far was to Harvard and to the Museum of Comparative Zoology—more than a million and a half dollars—but he earnestly hoped that that contribution would not be weighed in terms of dollars. In a letter to Sir John Murray he said: "While the sum total seems a large expenditure and one which appeals to the public and to the University officials, I hope that my influence on science at Cambridge will not always be measured by the dollar standard, as it is so apt to be. What I care for more is the recognition of the fact that having the means I have backed up my opinion of what was worth doing by a free expenditure of funds, and furthermore that I have since 1870 devoted my time as completely to the Museum as if I had been working on a salary of 1500 a year. And that since then I have published the results of my work continuously and hope to be judged by that and not by the total I may have spent for the same. I want to go down as a man of science and not to be known by a kind of cheap notoriety as an American millionaire." He need not have worried on that score. The scope and careful scholarship of his many publications would be enough to establish him as a man of science; beyond that, the very fiber of his mind reflects a quality of true greatness.

A legendary man indeed! And, as was fitting for a legendary figure, this one who had so loved the sea and its mysteries also died at sea. He had been touring Europe, and on his way home he stopped in London to dine with Sir John Murray and to book passage on the *Adriatic*. He sailed on March 23, 1910, and sometime early on Easter morning, March 27, he died quietly in his sleep. We could recall many tributes that the world of science paid to Alexander Agassiz and his work, but it seems more in keeping with Agassiz's universality of character to refer to Henry Adams, who, in *The Education of Henry Adams,* wrote of a world that was contemporary with Agassiz's, but very different from it. To H. L. Higginson shortly after Agassiz's death, Adams wrote:

"I wish I were there to show what respect I could for Alex. If I showed all I felt, it would be worth while to go far. He was the best we ever produced, and the only one of our generation whom I would have liked to envy. When I look back over our sixty years of life, and think of our millions of contemporaries, I am pacified when the figure of Alex occurs to me, and I feel almost reconciled to my own existence. We did one first-rate work when we produced him, and I do not know that, thus far, any other country has done better. I feel as though our lives had suddenly become poor—almost as though our generation were bankrupt by his loss. He stood so high above anyone else in my horizon that I can no longer see a landmark now that he is gone. To anyone else except you I should have to explain all this feeling, but you know how true and natural it is and I can leave it so."

12.
HERDMAN: OCEANOGRAPHY IN THE SERVICE OF MAN

As we have watched the progress of man's growing knowledge of the oceans from the ancient Polynesian voyagers to the nineteenth century's close, we have also followed the evolution of a new science. This was a far more complex science than any that had gone before, and it held out great promise for the benefit of mankind. It came to be called oceanography.

The classical sciences of the nineteenth century and earlier had their roots in mathematics, and as men began to apply mathematical concepts to the real world, they developed the physics of matter and energy and the chemistry of the elements and their compounds.

In marked contrast, men also studied nature at first hand, often without reference to the fundamental sciences at all. They followed, as it were, Louis Agassiz's dictum to "study nature, not books." That was reasonable enough in the early nineteenth century, for men still had much to learn that the intimate study of nature could tell them. This was generally known as "natural history," although at the time it was also called "science."

Natural history depended upon direct observation of whatever phenomena were thought to be of interest, and it was therefore essentially descriptive. However, the results of such observations could be subjected to inductive reasoning with enormously enlightening results, as Darwin showed through his theory (should we now call it law?) of natural selection.

Still, the method of observation, even when subjected to inductive reasoning, has serious limitations. Whenever we have reached a certain level of understanding of a thing, we always ask, "Then what would happen if . . . ?" In most cases nature does not provide the "if" situation, and we are forced to provide it artificially. This is the heart of the experimental method, which gives clear answers to "What would happen if . . . ?"

The sciences in which the experimental method gives the most reliable answers are those of physics and chemistry. Beyond them—as in the earth and the life sciences—the variables become harder and harder to know and control, but even there the method is very powerful.

As the nineteenth century progressed, a unified science of the sea began to emerge, although it would be more accurate to call it a discipline or branch of learning. Rather than being a science in its own right, it began to apply the insights of other sciences to the study of the ocean in all its aspects. Thus a problem in marine biology, for instance, might well be studied in the light of knowledge furnished by physicists, chemists, geologists, paleontologists, botanists, zoologists, ecologists, physiologists, biochemists, ethologists, embryologists, geneticists, and specialists in the developing theories of evolution.

One of the first to see the benefits that could follow the recognition of oceanography as a formal discipline was the Scottish marine biologist William A. Herdman. He studied zoology at the University of Edinburgh when Edinburgh was the world center of marine science, and became Wyville Thomson's assistant at the age of twenty-one. For two years he was closely associated with Thomson, John Murray, and other members of the *Challenger* staff, and he wrote the *Challenger* report on the tunicates, or sea squirts. In 1881 he went to the University of Liverpool, and he spent the rest of his life in teaching and in "interesting the public of Liverpool in the deeper knowledge of the seven seas that mean so much to that great port."

As a result of his nearly forty years of effort in university teaching and in educating the public, he was appointed First

Professor of Oceanography at the University of Liverpool. Well before that, however, he had begun to shape his thoughts on what a discipline of oceanography could become, and how it could contribute to the welfare of mankind. These thoughts he presented in an opening address before the British Association for the Advancement of Science at Ipswich in 1895. It is a pioneering paper that brings the natural history of the nineteenth century to a fitting close, and opens up new vistas for the ocean science of the twentieth.

OCEANOGRAPHY, BIONOMICS, AND AQUICULTURE

We include in our subject matter speciography and systematic zoology, which has been cultivated by the great classifiers and monographers from Linnaeus to Haeckel, and has culminated in our times in the magnificent series of fifty quarto volumes, setting forth the scientific results of the *Challenger* expedition; a voyage of discovery comparable only in its important and wide-reaching results with the voyages of Columbus, Gama, and Magellan at the end of the fifteenth century. It is now so long since the *Challenger* investigations commenced that few, I suppose, outside the range of professional zoologists are aware that although the expedition took place in 1872 to 1876, the work resulting therefrom has been going on actively until now—for nearly a quarter of a century in all—and in a sense, and a very real one, will never cease, for the *Challenger* has left an indelible mark upon science, and will remain through the ages exercising its powerful, guiding influence, like the work of Aristotle, Newton, and Darwin.

Most of the authors of the special memoirs on the sea and its various kinds of inhabitants have interpreted in a liberal spirit the instruction they received to examine and describe the collections entrusted to them, and have given us very valuable summaries of the condition of our knowledge of the animals in question, while some of the reports are little less than complete monographs of the groups. I desire to pay a tribute of respect to my former teacher and scientific chief, Sir Wyville Thomson, to whose initiative, along with Dr. W. B. Carpenter, we owe the first inception of our now celebrated deep-sea dredging expeditions, and to whose scientific enthusiasm, combined with administrative skill, is due in great part the successful accomplishment of the *Lightning,* the *Porcupine,* and the *Challenger* expeditions. Wyville Thomson lived long enough to superintend the first examination of the collections brought home, their division into groups, and the allotment of these to specialists for description. He enlisted the services of his many scientific friends at home and abroad, he arranged the general plan of the work, decided upon the form of publication, and died in 1882, after seeing the first ten or twelve zoological reports through the press.

Within the last few months have been issued the two concluding volumes of this noble series, dealing with a summary of the results, conceived and written in a masterly manner by the eminent editor of the reports, Dr. John Murray. An event of such first-rate importance in zoology as the completion of this great work ought not to pass unnoticed at this zoological gathering. I desire to express my appreciation and admiration of Dr. Murray's work, and I do not doubt that the section will permit me to convey to Dr. Murray the congratulations of the zoologists present, and their thanks for his splendid services to science. Murray, in these "Summary" volumes, has given definiteness of scope and purpose and a tremendous impulse to that branch of science—mainly zoological—which is coming to be called oceanography.

OCEANOGRAPHY

Oceanography is the meeting ground of most of the sciences. It deals with botany and zoology, "including animal physiology"; chemistry, physics, mechanics, meteorology, and geology all contribute, and the subject is of course intimately connected with geography, and has an incalculable influence upon mankind, his distribution, characteristics, commerce, and economics. Thus oceanography, one of the latest developments of marine zoology, extends into the domain of, and ought to find a place in, every one of the sections of the British Association.

Along with the intense specialization of certain lines of zoology in the last quarter of the nineteenth century, it is important to notice that there are also lines of investigation which require an extended knowledge of, or at least make use of the results obtained from various distinct subjects. One of these is oceanography, another is bionomics, which I have referred to above, a third is the philosophy of zoology, or all those studies which bear upon the theory of evolution, and a fourth is the investigation of practical fishery problems, which is chiefly an application of marine zoology. Of these four subjects—which, while analytical enough in the detailed investigation of any particular problem, are synthetic in drawing together and making use of the various divergent branches of zoology and the neighboring sciences—oceanography, bionomics, and the fisheries investigation are most closely related, and I desire to devote the remainder of this address to the consideration of some points in connection with their present position.

Dr. Murray, in a few only too brief paragraphs at the end of his detailed summary of the results of the *Challenger* expedition, which I have alluded to above, states some of the views, highly suggestive and original, at which he has himself arrived from his unique experience. Some of his conclusions are very valuable contributions to knowledge, which will no doubt be adopted by marine zoologists. Others, I venture to think, are less sound and well founded, and will scarcely stand the test of time and further experience. But for all such statements, or even suggestions, we should be thankful. They do much to stimulate further research; they serve, if they can neither be refuted nor established, as working hypotheses; and even if they have to be eventually abandoned, we should bear in mind what Darwin has said as to the difference in their influence on science between erroneous facts and erroneous theories: "False facts are highly injurious to the progress of science, for they often endure long; but false views, if supported by some evidence, do little harm, for everyone takes a salutary pleasure in proving their falseness, and when this is done, one path toward error is closed, and the road to truth is often at the same time opened." (Darwin: *The Descent of Man,* second edition, 1882, p. 606.)

Probably no group of animals in the sea is of so much importance from the point of view of food as the Copepoda. They form a great part of the food of whales, and of herrings and many other useful fish, both in the adult and in the larval state, as well as of innumerable other animals, large and small. Consequently, I have inquired somewhat carefully into their distribution in the sea, with the assistance of Professor Brady, Mr. Scott, and Mr. Thompson. These experienced collectors all agree that Copepoda are most abundant both as to species and individuals, close round the shore, amongst seaweeds, or in shallow water in the Laminarian zone over a weedy bottom. Individuals are sometimes extremely abundant on the surface of the sea amongst the plankton, or in shore pools near high water, where, amongst *Enteromorpha*, they swarm in immense profusion; but for a gathering rich in individuals, species and genera, the experienced collector goes to the shallow waters of the Laminarian zone. In regard to the remaining, higher groups of the Crustacea, my friend Mr. Alfred O. Walker tells me that he considers them most abundant at depths of from 0 to 20 fathoms.

I hope no one will think that these are detailed matters interesting only to the collector, and having no particular bearing upon the great problems of biology. The sea is admittedly the starting point of life on this earth, and the conclusions we come to as to the distribution of life in the different zones must form and modify our views as to the origin of the faunas—as to the peopling of the deep sea, the shallow waters, and the land. Murray supposes that life started in Pre-Cambrian times on the mud, and from there spread upward into shallower waters, outward on to the surface, and, a good deal later, downward to the abysses by means of the cold Polar waters. The late Professor Moseley considered the pelagic, or surface life of the ocean, to be the primitive life from which all the others have been derived. Professor W. K. Brooks (*The Genus Salpa*, 1893, p. 156, etc.) considers that there was a primitive pelagic fauna, consisting of the simplest microscopic plants and animals, and that "pelagic life was abundant for a long period during which the bottom was uninhabited."

I consider that the Laminarian zone close to low-water mark is at present the richest in life, that it probably has been so in the past, and that if one has to express a more definite opinion as to where, in Pre-Cambrian times, life in its simplest forms first appeared, I see no reason why any other zone should be considered as having a better claim than what is now the Laminarian to this distinction. It is there, at present, at any rate, in the upper edge of the Laminarian zone, at the point of junction of sea, land, and air, where there is a profusion of food, where the materials brought down by streams or worn away from the land are first deposited, where the animals are able to receive the greatest amount of light and heat, oxygen and food, without being exposed periodically to the air, rain, frost, sun, and other adverse conditions of the Littoral zone. It is there that life—it seems to me—is most abundant, growth most active, competition most severe. It is there, probably, that the surrounding conditions are most favourable to animal life; and, therefore, it seems likely that it is from this region that, as the result of overcrowding, migrations have taken place downward to the abysses, outward on the surface, and upward on to the shore. Finally, it is in this Laminarian zone, probably, that under the stress of competition between individuals and between allied species evolution of new forms by means of natural selection has been most active. Here, at any rate, we find, along with some of the most primitive of animals, some of the most remarkably modified forms, and some of the most curious cases of minute adaptation to environment. This brings us to the subject of

BIONOMICS,

which deals with the habits and variations of animals, their modifications, and the relations of these modifications to the surrounding conditions of existence.

It is remarkable that the great impetus given by Darwin's work to biological investigation has been chiefly directed to problems of structure and development, and not so much to bionomics until lately. Variations amongst animals in a state of nature, is however at last beginning to receive the attention it deserves. Bateson has collected together, and classified in a most useful book of reference, the numerous scattered observations on variation made by many investigators, and has drawn from some of these cases a conclusion in regard to the discontinuity of variation which many field zoologists find it hard to accept.

Weldon and Karl Pearson have recently applied the methods of statistics and mathematics to the study of individual variation. This method of investigation, in Professor Weldon's hands, may be expected to yield results of great interest in regard to the influence of variations in the young animal upon the chance of survival, and so upon the adult characteristics of the species. But while acknowledging the value of these methods, and admiring the skill and care with which they have been devised and applied, I must emphatically protest against the idea which has been suggested, that only by such mathematical and statistical methods of study can we successfully determine the influence of the environment on species, gauge the utility of specific characters, and throw further light upon the origin of species. For my part, I believe we shall gain a truer insight into these mysteries which still involve variations and species by a study of the characteristic features of individuals, varieties, and species in a living state in relation to their environment and habits.

The mode of work of the old field naturalists, supplemented by the apparatus and methods of the modern laboratory, is, I believe, not only one of the most fascinating, but also one of the most profitable fields of investigation for the philosophical zoologist. Such studies must be made in that modern outcome of the growing needs of our science, the Zoological Station, where marine animals can be kept in captivity under natural conditions, so that their habits may be closely observed, and where we can follow out the old precept—first, observation and reflection; then experiment.

The biological stations of the present day represent, then, a happy union of the field work of the older naturalists with the laboratory work of the comparative anatomist, histologist, and embryologist. They are the culmination of the "Aquarium" studies of Kingsley and Gosse, and of the feeling in both scientific men and amateurs, which was expressed by Herbert Spencer when he said, "Whoever at the seaside has not had a microscope and an aquarium has yet to learn what the highest pleasures of the seaside are." Moreover, I feel that the biological station has come to the rescue, at a critical moment, of our laboratory worker who, without its healthy, refreshing influence, is often, in these latter days, in peril of losing his intellectual life in the weary maze of microtome methods and transcendental cytology. The old Greek myth of the Libyan giant, Antaeus, who wrestled with Hercules and regained his strength each time he touched his mother earth, is true at least of the zoologist.

I am sure he derives fresh vigor from every direct contact with living nature.

In our tanks and artificial pools we can reproduce the Littoral and the Laminarian zones; we can see the methods of feeding and breeding—the two most powerful factors in influencing an animal. We can study mimicry, and test theories of protective and warning coloration.

The explanations given by these theories of the varied forms and colors of animals were first applied by such leaders in our science as Bates, Wallace, and Darwin, chiefly to insects and birds, but have lately been extended, by the investigations of Giard, Garstang, Clubb, and others, to the case of marine animals. I may mention very briefly one or two examples. Amongst the Nudibranchiate Mollusca—familiar animals around most parts of our British coasts—we meet with various forms which are edible, and, so far as we know, unprotected by any defensive or offensive apparatus. Such forms are usually shaped or colored so as to resemble more or less their surroundings, and so become inconspicuous in their natural haunts. *Dendronotus arborescena,* one of the largest and most handsome of our British Nudibranches, is such a case. The large, branched processes on its back, and its rich purple-brown and yellow markings, tone in so well with the masses of brown and yellow zoophytes and purplish red seaweeds, amongst which we usually find *Dendronotus,* that it becomes very completely protected from observation, and, as I know from my own experience, the practiced eye of the naturalist may fail to detect it lying before him in the tangled forests of a shore pool.

Other Nudibranches, however, belonging to the genus *Eolis,* for example, are colored in such a brilliant and seemingly crude manner that they do not tone in with any natural surroundings, and so are always conspicuous. They are active in their habits, and seem rather to court observation than to shun it. When we remember that such species of *Eolis* are protected by the numerous stinging cells in the edinophorous sacs placed on the tips of all the dorsal processes, and that they do not seem to be eaten by other animals, we have at once an explanation of their fearless habits and of their conspicuous appearance. The brilliant colors are in this case of a warning nature, for the purpose of rendering the animal provided with the stinging cells noticeable and recognizable. But it must be remembered that in a museum jar, or in a laboratory dish, or as an illustration in a book or on a wall, *Dendronotus* is quite as conspicuous and striking an animal as *Eolis*. In order to interpret correctly the effect of their forms and colors we must see them alive and at home, and we must experiment upon their edibility or otherwise in the tanks of our biological stations.

Let me give you one more example of a somewhat different kind. The soft, unprotected mollusc, *Lamellaria perspicua,* is not uncommonly found associated (as Giard first pointed out) with colonies of the compound Ascidian *Leptoclinum maculatum,* and in these cases the *Lamellaria* is found to be eating the *Leptoclinum,* and lies in a slight cavity which it has excavated in the Ascidian colony, so as to be about flush with the general surface. The integument of the mollusc is, both in general tint, and also in surface markings, very like the Ascidian colony with its scattered ascidiozoids [the individuals of a compound ascidian]. This is clearly a good case of protective coloring. Presumably, the *Lamellaria* escapes the observation of its enemies through being mistaken for a part of the *Leptoclinum* colony; and the *Leptoclinum* being crowded like a sponge with minute, sharp-pointed spicules is, I suppose, avoided as inedible by carnivorous animals, which might devour such things as the soft, unprotected mollusc. But the presence of the spicules evidently does not protect the *Leptoclinum* from *Lamellaria,* so that we have, if the above interpretation is correct, the curious result that the *Lamellaria* profits by a protective characteristic of the *Leptoclinum,* for which it has itself no respect, or to put it another way, the *Leptoclinum* is protected against enemies to some extent for the benefit of the *Lamellaria,* which preys upon its vitals.

It is to my mind no sufficient objection to theories of protective coloration that careful investigation may from time to time reveal cases where a disguise is penetrated, a protection frustrated, an offensive device supposed to confer inedibility apparently ignored. We must bear in mind that the enemies, as well as their prey, are exposed to competition, are subject to natural selection, are undergoing evolution, that the pursuers and the pursued, the eaters and the eaten, have been evolved together, and that it may be of great advantage to be protected from some even if not from all enemies. Just as on land, some animals can browse upon thistles whose "nemo me impune lacessit" spines are supposed to confer immunity from attack, so it is quite in accord with our ideas of evolution by means of natural selection to suppose that some marine animals have evolved an indifference to the noxious sponge or the bristling Ascidian, which are able, by their defensive characteristics, like the thistle, to repel the majority of invaders.

Although we can keep and study the Littoral and Laminarian animals at ease in our zoological stations, it may perhaps be questioned how far we can reproduce in our experimental and observational tanks the conditions of the Coralline and the Deep-mud zones. One might suppose that the pressure—which we have no means as yet for supplying—and which at 30 fathoms amounts to nearly 100 pounds to the square inch, at 80 fathoms to about 240 pounds, or over 2 hundred-weight on the square inch, would be an essential factor in the life conditions of the inhabitants of such depths, yet we have kept half a dozen specimens of *Calocaris macandreae,* dredged from 70 to 80 fathoms, alive at the Port Erin Biological Station for several weeks; we have had both the red and yellow forms of *Sarcodictyon catenata,* dredged from 30 to 40 fathoms, in a healthy condition with the polyps freely expanded for an indefinite period; and Mr. Arnold Watson has kept the Polynoid worm, *Panthalis oerstedi* from the deep mud at over 50 fathoms, alive, healthy, and building its tube under observation, first for a week at the Port Erin Station, and for many months at Sheffield, in a comparatively small tank with no depth of water. Consequently it seems clear that with ordinary care almost any marine animals from such depths as are found within the British area may be kept under observation and submitted to experiment in healthy and fairly natural conditions. The biological station, with its tanks, is in fact an arrangement whereby we bring a portion of the sea with its rocks and bottom deposits and seaweeds, with its inhabitants and their associates, their food and their enemies, and place it for continuous study on our laboratory table. It enables us to carry on the bionomical investigations to which we look for information as to the methods and progress of evolution; in it lie centered our hopes of a comparative physiology of the invertebrates—a physiology not wholly medical—and finally to the biological station we confidently look for help in connection with our coast fisheries. This brings me to the last subject which I shall touch upon, a subject closely related both to oceanography and bionomics, and one which depends much for its future advance upon our biological stations—that is the subject of

AQUICULTURE

or industrial ichthyology, the scientific treatment of fishery investigations, a subject to which Professor McIntosh has first in this country directed the attention of zoologists, and in which he has been guiding us for the last decade in his admirable researches. What chemistry is to the aniline, the alkali, and some other manufactures, marine zoology is to our fishing industries.

Although zoology has never appealed to popular estimation as a directly useful science having industrial applications in the same way that chemistry and physics have done, and consequently has never had its claims as a subject of technical education sufficiently recognized, still, as we in this section are well aware, our subject has many technical applications to the arts and industries. Biological principles dominate medicine and surgery. Bacteriology, brewing, and many allied subjects are based upon the study of microscopic organisms. Economic entomology is making its value felt in agriculture. Along all these and other lines there is a great future opening up before biology, a future of extended usefulness, of popular appreciation, and of value to the nation—and not the least of these technical applications will, I am convinced, be that of zoology in our fishing industries. When we consider their enormous annual value—about eight millions sterling at first hand to the fisherman, and a great deal more than that by the time the products reach the British public, when we remember the very large proportion of our population who make their living directly or indirectly (as boat builders, net makers, etc.) from the fisheries, and the still larger proportion who depend for an important element in their food supply upon these industries, when we think of what we pay other countries—France, Holland, Norway—for oysters, mussels, lobsters, etc., which we could rear in this country if our sea shores and our sea bottom were properly cultivated; and when we remember that fishery cultivation or aquiculture is applied zoology, we can readily realize the enormous value to the nation which this direct application of our science will one day have—perhaps I ought rather to say, we can scarcely realize the extent to which zoology may be made the guiding science of a great national industry.

The flourishing shellfish industries of France, the oyster culture at Arachon and Marennes, and mussel culture by bouchots [long irregular arrays of vertical stakes interlaced with twigs] in the Bay of Aiguillon, show what can be done as the result of encouragement and wise assistance by the Government, with constant industry on the part of the people, directed by scientific knowledge. In another direction the successful hatching of large numbers (hundreds of millions) of cod and plaice by Captain Dannevig in Norway, and by the Scottish fishery board at Dunbar, opens up possibilities of immense practical value in the way of restocking our exhausted bays and fishing banks, depleted by the overtrawling of the last few decades.

The demand for the produce of our sea is very great, and would probably pay well for an increased supply. Our choice fish and shellfish are becoming rarer and the market prices are rising. The great majority of our oysters are imported from France, Holland, and America. Even in mussels we are far from being able to meet the demand. In Scotland alone the long-line fishermen use nearly a hundred millions of mussels to bait their hooks every time the lines are set, and they have to import annually many tons of these mussels at a cost of from £3 to £3 10s. a ton.

Whether the wholesale introduction of the French method of mussel culture, by means of bouchots, on to our shores would be a financial success is doubtful. Material and labor are dearer here, and beds, scars, or scalps seem, on the whole, better fitted to our local conditions; but as innumerable young mussels all around our coasts perish miserably every year for want of suitable objects to attach to, there can be no reasonable doubt that the judicious erection of simple stakes or plain bouchots would serve a useful purpose, at any rate in the collection of seed, even if the further rearing be carried out by means of the bed system.

All such aquicultural processes require, however, in addition to the scientific knowledge, sufficient capital. They can not be successfully carried out on a small scale. When the zoologist has once shown, as a laboratory experiment in the zoological station, that a particular thing can be done—that this fish can be hatched or that shellfish reared—under certain conditions which promise to be an industrial success, then the matter should be carried out by the Government or by capitalists on a sufficiently huge scale to remove the risk of results being vitiated by temporary accident or local variation in the conditions. It is contrary, however, to our English traditions for Government to help in such a matter, and if our local sea fisheries committees have not the necessary powers nor the available funds, there remains a splendid opportunity for opulent landowners to erect sea-fish hatcheries on the shores of their estates, and for the rich merchants of our great cities to establish aquiculture in their neighboring estuaries, and by so doing instruct the fishing population, resuscitate the declining industries, and cultivate the dying shores—in all reasonable probability to their own profit.

In addition to the farming of our shores, there is a great deal to be done in promoting the fishing industries on the inshore and offshore grounds along our coast, and in connection with such work the first necessity is a thorough scientific exploration of our British seas by means of a completely fitted dredging and trawling expedition. Such exploration can only be done in little bits, spasmodically, by private enterprise. From the time of Edward Forbes it has been the delight of British marine zoologists to explore, by means of dredging from yachts or hired vessels during their holidays, whatever areas of the neighboring seas were open to them. Some of the greatest names in the roll of our zoologists, and some of the most creditable work in British zoology, will always be associated with dredging expeditions. Forbes, Wyville Thomson, Gwyn Jeffreys, McIntosh, and Norman—one can scarcely think of them without recalling

> *Hurrah for the dredge, with its iron edge,*
>   *And its mystical triangle.*
> *And its hided net, with meshes set,*
>   *Odd fishes to entangle!*[1]

Much good pioneer work in exploration has been done in the past by these and other naturalists, and much is now being done locally by committees or associations—by the Dublin

---

[1] From Edward Forbes' "Song of the Dredge"; see Chapter 5, above.

Royal Society on the west of Ireland, by the Marine Biological Association at Plymouth, by the Fisheries Board in Scotland, and by the Liverpool Marine Biology Committee in the Irish Sea; but few zoologists or zoological committees have the means, the opportunity, the time to devote, along with their professional duties, to that detailed, systematic survey of our whole British sea area which is really required. Those who have had no experience of it can scarcely realize how much time, energy, and money is required to keep up a series of dredging expeditions, how many delays, disappointments, expensive accidents, and real hardships there are, and how often the naturalist is tempted to leave unprofitable grounds, which ought to be carefully worked over, for some more favored spot where he knows he can count on good spoil. And yet it is very necessary that the whole ground—good or bad though it may be from a zoological point of view—should be thoroughly surveyed, physically and biologically, in order that we may know the conditions of existence which environ our fishes on their feeding grounds, their spawning grounds, their "nurseries," or whatever they may be.

The British Government has done a noble piece of work, which will redound to its everlasting credit, in providing for and carrying out the *Challenger* expedition. Now that that great enterprise is completed, and that the whole scientific world is united in appreciation of the results obtained, it would be a glorious consequence, and surely a very wise action in the interests of the national fisheries, for the Government to fit out an expedition, in charge of two or three zoologists and fisheries experts, to spend a couple of years in exploring more systematically than has yet been done, or can otherwise be done, our British coasts from the Laminarian zone down to the deep mud. No one could be better fitted to organize and direct such an expedition than Dr. John Murray.

Such a detailed survey of the bottom and the surface waters, of their conditions and their contents, at all times of the year for a couple of years, would give us the kind of information we require for the solution of some of the more difficult fisheries problems—such as the extent and causes of the wanderings of our fishes, which "nurseries" are supplied by particular spawning grounds, the reason of the sudden disappearance of a fish, such as the haddock, from a locality, and in general the history of our food fishes throughout the year.

It is creditable to our Government to have done the pioneer work in exploring the great ocean, but surely it would be at least equally creditable to them—and perhaps more directly and immediately profitable, if they look for some such return from scientific work—to explore our own seas and our own sea fisheries.

There is still another subject connected with the fisheries which the biologist can do much to elucidate—I mean the diseases of edible animals and the effect upon man of the various diseased conditions. It is well known that the consumption of mussels taken from stagnant or impure water is sometimes followed by severe symptoms of irritant poisoning which may result in rapid death. This "musselling" is due to the presence of an organic alkaloid or ptomaine, in the liver of the mollusc, formed doubtless by a microorganism in the impure water. It is clearly of the greatest importance to determine accurately under what conditions the mussel can be infected by the microorganism, in what stage it is injurious to man, and whether, as is supposed, steeping in pure water with or without the addition of carbonate of soda will render poisonous mussels fit for food.

During the last year there has been an outcry, almost amounting to a scare, and seriously affecting the market,[2] as to the supposed connection between oysters taken from contaminated water and typhoid fever. This, like the musselling, is clearly a case for scientific investigation, and, with my colleague, Professor Boyce, I have commenced a series of experiments and observations, partly at the Port Erin Biological Station, where we have oysters laid down on different parts of the shore under very different conditions, as well as in dishes and tanks, and partly at University College, Liverpool.

Our object is to determine the effect of various conditions of water and bottom upon the life and health of the oyster, the effect of the addition of various impurities to the water, the conditions under which the oyster becomes infected with the typhoid bacillus, and the resulting effect upon the oyster, the period during which the oyster remains infectious, and, lastly, whether any simple practicable measures can be taken (1) to determine whether an oyster is infected

---

[2] I am told that between December and March the oyster trade decreased 75 per cent.

with typhoid, and (2) to render such an oyster innocuous to man. As Professor Boyce and I propose to lay a paper upon this subject before the section, I shall not occupy further time now by a statement of our methods and results.

I have probably already sufficiently indicated to you the extent and importance of the applications of our science to practical questions connected with our fishing industries. But if the zoologist has great opportunities for usefulness, he ought always to bear in mind that he has grave responsibilities in connection with fisheries investigations. Much depends upon the results of his work. Private enterprise, public opinion, local regulations, and even imperial legislation, may all be affected by his decisions. He ought not lightly to come to conclusions about weighty matters. I am convinced that of all the varied lines of research in modern zoology, none contains problems more interesting and intricate than those of bionomics, oceanography, and the fisheries, and of these three series the problems connected with our fisheries are certainly not the least interesting, not the least intricate, and not the least important in their bearing upon the welfare of mankind.

Especially in view of the fact that Sir William A. Herdman delivered his address as long ago as 1895, we must recognize it as a truly remarkable document. The substance of his argument is as valid today as it was three-quarters of a century ago. Oceanography holds out an even greater hope for the benefit of mankind; the term "bionomics" has been largely replaced by "ecology," but the discipline itself has increased in interest, intricacy, and promise; and since some two billion human beings have been added to the world's population since Sir William spoke, new knowledge of how to conserve and increase the food productivity of the ocean is of correspondingly critical importance.

Quite naturally Sir William spoke of the problems of the British seas. He would be pleased if he could now see how far his philosophy has been extended to the oceans of the world. Although the cost of the surveys and researches that he envisioned have risen phenomenally, government and private support of them—both nationally and internationally—has been substantial. The seaside laboratories of the world have burgeoned, and their contributions both to basic science and to practical technology have been great. How great, we shall see in selected cases in the chapters that follow.

13.
HJORT AND HENSEN: THE DYNAMICS OF OCEAN POPULATIONS

"In a story which they relate of a most unfortunate king, the Norwegian sagas tell us that during his reign famine and distress were rampant in the land. One autumn, to propitiate the gods, oxen were sacrificed, but the following year the crops proved no better and men instead of oxen were chosen for the sacrifice. Even this, however, failed to placate the gods, and accordingly in the third autumn the chief men among the people agreed that their king was the cause of the famine. They therefore seized the king and killed him, that they might stain the altars of the gods with his blood. And the good years returned and there was peace and plenty in the land. The fairy tales of our own times, a thousand years after this episode, differ from the story of the sagas. But do they represent an intellectual advance over that story?"

With this question the Norwegian biologist Johan Hjort opened a series of lectures that he gave at the University of Oslo in the spring of 1936. He was then approaching seventy, he had been honored throughout the world of science for his research on fisheries problems, and that fall he joined an international gathering at Harvard's Museum of Comparative Zoology, where he saw his lecture series into print in English under the title *The Human Value of Biology*.

Hjort wrote of his early concern about the good and bad years that Norwegian fishermen experienced, about how that had led him to seek out the reasons for lush and lean years in farming, and about how an understanding of the fluctuations of human populations might throw light on both of these. He closed the book with some discerning observations on what biology can teach us for the better conduct of human affairs.

For many years Hjort's principal interest lay in trying to find out why the cod and herring catches in northern European waters were so widely variable, for then, as now, the herring and the cod were the principal food fishes of the world.[1] Well before receiving his Ph.D. at the University of Munich in 1902, he was fascinated by the problems of marine biology, and as we saw in an earlier chapter, it was he who fitted out the Norwegian research vessel *Michael Sars* for scientific work in 1900. Soon after, the Norwegian cod fisheries had a succession of very lean years, and Hjort and his team of scientists assigned to the *Michael Sars* began an intensive investigation of the reasons for the decline. They worked on the problem for more than a decade, and their far-reaching conclusions were presented in Hjort's milestone paper of 1914, "Fluctuations in the Great Fisheries of Northern Europe." Although today these conclusions must be considered somewhat oversimplified, they nevertheless represented a brilliant advance sixty years ago.

Then, as now, popular opinion tended to assume that overfishing was the basic cause of any decline in the productivity of the fisheries. Hjort and his colleagues were not willing to accept such an easy answer. They insisted on finding out what was really going on, and they began by studying the distributions of fishes according to their ages, in samples taken at many different times and places. If, for example, they should find an exceptionally large number of five-year-old fish in a certain place, they could then ask what had happened five years ago to cause that. Conversely, if three-year-old fish were few, they could ask what had happened three years earlier.

But how does one tell the age of a fish? Hjort first thought

---

[1] In recent years the Peruvian anchovetas have been taken in greater tonnages than herring and cod combined, but anchovetas are mostly ground into meal for feeding chickens, not used directly as human food.

of weighing them, but he soon learned that the weight of a fish tells no more about its age than the weight of a man tells about his.[2] Hjort had heard that marine biologists on the German island of Helgoland were dating fish by annual growth rings, like those of a tree, or of the corals that we saw earlier. In fishes, these rings are found both in the scales and in the otoliths, which are calcareous growths located in a fish's ear. Refining these studies, Hjort found that the annual rings in the scales were convenient and highly reliable for dating herring and cod, but that the scales of the plaice (a flatfish like a flounder) were too small and delicate to be manageable. For plaice, the otoliths served well.

Somewhat later the Norwegians Einar Lea and Per Ottestad made a more exhaustive study of scale dating; they fully confirmed its accuracy, and found further that certain relationships among the annual rings revealed what part of the North Atlantic a fish came from and even the temperature of the water in which it swam.

Now Hjort and his staff of the *Michael Sars* proceeded, year after year, to amass data on the ages of herring and cod in many different areas. What they found was—even to them—most unexpected.

One might guess that the distributions of fish populations according to age would be similar from one year to the next; even though one year should yield many fish and the next year only a few, we would hardly suppose that the age distributions would be very different. This is what Hjort expected, and in that he was influenced by his continuing interest in the statistics of human populations. As early as 1907 he gave a now-famous lecture, which was translated into English and published in Edinburgh in 1908. There he said:

"I will proceed to draw a comparison between this Fishery Research and a science which is much more generally understood. I mean the science of Vital Statistics.

"In all the expositions of the science of vital statistics there are three prominent features which attract our chief consideration: 1. Birth rate. 2. Age distribution. 3. Migration. It is customary to study these questions by the help of what are called representative

---

[2] In an outrageous spoof that Kenneth Roberts wrote for the *Boston Post* at about this time, he noted that "unlike the human race, the cod weighs more when undressed than when dressed." In *I Wanted to Write*, Roberts reported a series of fishy interviews with one Professor Morton Kilgallen, F.R.S., F.R.G.S., of Balliol College, who claimed to "tell the age of a cod by the length of its whisker, even though slightly mossy around the edges."

statistics. A certain number of individuals are selected that are supposed to stand for the mass of the people, and attention is directed to them. We ascertain from this source their average length of life, their wanderings, their increase or decrease, and whether sickness, war, disaster, or emigration play any appreciable part in reducing the population.

"It seems at first sight a bold suggestion to propose the study of the fish supply along lines like these. A population can be counted; but who knows how many fishes are in the sea? And yet it appears to me a project great in possibilities, to regard the discoveries of fishery research from a similar standpoint to what has been adopted in the science of vital statistics."

This graph, called a histogram, shows how the weights of 120 students were found to be distributed. Although it is skewed somewhat to the left, the distribution approximates the theoretical "normal distribution" of the statisticians.

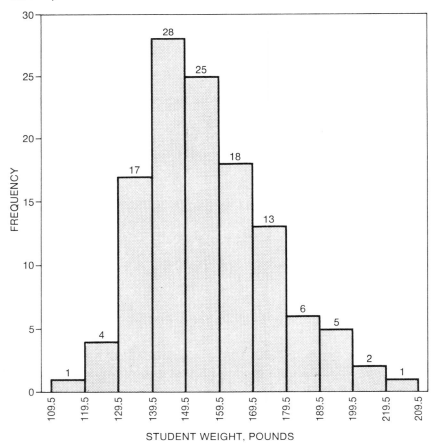

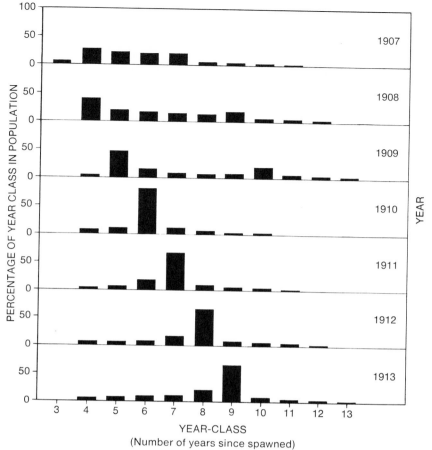

Seven histograms indicating the age distribution of herring for the years 1907 through 1913. The heights of the bars show the percentages of fish found in each year-class. In 1908, for example, the fish in year-class 4 (which were spawned in 1904) contributed some 37 per cent of the sample. In 1909, the 1904 fish were in year-class 5, and constituted some 45 per cent of the sample. In all succeeding years the brood of 1904 continued to predominate.

How "great in possibilities" Hjort's idea was can be strikingly shown by a simple comparison. Let us look first at a typical statistical study of a sample of a human population, and then at Hjort's own statistics on the age distribution of a number of samples of herring.

Suppose that we wish to know how the weights of the individuals are distributed among a class of students in a dormitory. We can easily record the weight of each, arrange the numbers in a statistical table, and display them in graphic form on a histogram. Such a histogram is shown in the accompanying figure, from which we see that among 120 students there were a few lightweights, a few heavyweights, and that most of the population clustered around a median or central value. Although the sample is skewed somewhat to the left, its bell-shaped form is quite typical of many parameters of human population samples.

Would we then expect that the ages of the individuals in a sample of a herring population would be distributed in more or less the same way? The next figure shows what Hjort actually found in the Norwegian herring shoals from 1907 to 1913.[3] What springs to our attention is the fact that the fish that were spawned in 1904 were tremendously more abundant than those spawned in any other year, and that this situation continued for years. In extreme cases, the year-class of 1904 exceeded the productivity of other years by as much as thirty to fifty times.

With this irregularity well established, Hjort's next problem was to find the reason for it. A number of different possibilities came to mind. One was that 1904 could have been an exceptionally good spawning year, with the production of vast numbers of eggs, while in other years fewer eggs might have been produced. On the face of it, this was a dubious guess, for mature fish produce eggs in astounding numbers. For example, a single large cod may produce as many as four million eggs, and only a tiny fraction of these would need to survive in order to produce a record year.

Hjort also considered the effects of storms and of variations in water temperatures from year to year, but these conjectures were inconclusive, and he began to suspect that the critical period in a fish's life occurred during the few short weeks after its egg was hatched. When a larval fish first emerges, it looks like nothing more than a fine black thread attached to a globular egg much larger than itself. As the days pass, the larval fish slowly absorbs the yolk of the egg and begins to look more and more like a little fish. During this period the young fish starts to take some food by mouth, and of course, when it has used up the yolk, it is entirely dependent on foraging in the open water.

The general nature of fingerling food was well known at that time. Most of it consisted of microscopic one-celled plants, and most of these were the beautiful diatoms that have long been the delight of amateur microscopists. These range in size from one three-thousandth of an inch to one-thirtieth of an inch, the larger

---

[3] For this graph I have used the data of Dr. Einar Lea, since they are somewhat more regular than those first found by Hjort.

ones floating free as separate individuals, the smaller ones tending to aggregate in chains. In lesser numbers there are the peridinians, with shells of cellulose rather than silica, and coccolithophores with protective disks of lime. All of these organisms contain green chlorophyll, which enables them to manufacture carbohydrates from carbon dioxide and water, converting the energy of sunlight into stored chemical energy for use in their own life processes and for consumption by the small animals that also float freely in the sea. A large population of these plants is required to sustain animals, for we would ordinarily find about one hundred and fifty times as many plants as animals in a typical sample of sea water. Collectively, these and other free-floating plants and animals are called plankton, after the Greek *planktós*, "drifting," and they constitute the primary source of food for all the free-swimming animals, from herring larvae to whales.

Since Johan Hjort was already deeply impressed by the extreme variability of fish populations, he quite naturally suspected that plankton populations would also vary widely, and if that proved to be the case, it would provide a clue to the good and poor year-classes of fish.

Therefore, in the spring of 1913 he went to the Lofoten fishing grounds to see if he could establish a connection between the abundance of the plankton and the survival of fish larvae. Using fine silk nets, he took sample after sample over many weeks, and discovered first that there were wide differences in the composition of his catches. His first hauls contained almost nothing but fish eggs, and he was sure that if these hatched when no other food were available, the emergent fish would starve. Then as the weeks passed, the numbers of plankton increased, and "later on in the spring . . . enormous quantities of microscopic plant organisms [made] their appearance"; obviously larvae that hatched at that time would be well fed. Hjort's conclusion was that good years for plankton meant good years for fish, and vice versa.

That, however, was only one step closer to the real answer, for it left the tantalizing question: "Why should the numbers of plankton vary so widely?" Hjort turned hopefully to the man who by reputation was the world authority on plankton, Viktor Hensen, who had named this great class of organisms as early as 1887.

Hensen was a professor of physiology at the University of Kiel in northern Germany. Some thirty-five years Hjort's senior, Hensen had had a curiously varied career. He studied science and medicine at the universities of Würzburg, Berlin, and Kiel, and after he took his degree from Kiel, he became an assistant in the Department of Physiology and did his early research on embryology and histology. He was appointed to the chair of physiology in Kiel, and studied how the body handles sugar; independent of Claude Bernard, he found that sugar is stored in the form of a starch called glycogen, and that glycogen breaks down to release glucose into the blood stream.

Hensen also showed that the lens of the eye recedes when the eye accommodates, and made important contributions to the understanding of the sense of touch and of the physiology of hearing and of reproduction. The latter part of his life he devoted to marine biology, especially to the study of the microscopic plants in the planktonic community. He was a member of the German Plankton Expedition of 1889, and he edited the reports of that voyage.

Hensen became obsessed with an effort to estimate the total mass of living creatures in the sea—the biomass—and he made an important step toward that goal with his invention of new and more refined plankton nets. These were designed in definite shapes and dimensions intended to strain a column of water of specified volume "and give a catch which, when multiplied by a coefficient, would be the exact contents of, say, a square meter in section—a most desirable result," as Sir William A. Herdman wryly commented in *Founders of Oceanography and Their Work*, "if possible of attainment."

Hensen hoped that it would be possible, and he went so far as to estimate the total biomass of the world's oceans. From his hauls merely in the North Sea and the southern Baltic he calculated that, in microscopic plants alone, the ocean was as productive as an equal area of prairie land.

What was especially notable was his finding that no significant numbers of plants were encountered at depths greater than two hundred meters, which is about the maximum depth that light can penetrate sea water. But even though the basic mechanism of photosynthesis had long been known, and even though Hensen was a physiologist, he does not seem to have appreciated the indispensable role that sunlight plays in support of all living organisms in the sea.

Hensen's estimates of the ocean's biomass were vulnerable to the most serious criticism, for he carried his conclusions far beyond the reliability of his data. He assumed that populations were uniformly distributed over wide areas, and even at the time of the *Challenger* expedition that was known not to be so. He assumed that populations were uniform over time, and even his own data should have told him that that was not so. Finally, he assumed that a comparatively small number of samples would be statistically significant, and this was at least open to serious question. In spite of these faults, however, Hensen did arouse widespread interest, both among scientists and among the general public, in the study of planktonic plants and animals, and he paved the way for the more advanced work that was to follow. He was, however, unable to tell Hjort why the planktonic plants varied as much as they did. Indeed, as we shall see, that was not to be learned until much later.

Even today it is hazardous to give more than the roughest estimates for the total biomass in the sea. But we can at least get some impression of its great size from a simple example. The blue whale (for our example) eats only one kind of food, the euphausians, the shrimplike animals, some two inches long, that whalers call "krill."

One blue whale weighs about 100 tons, and to maintain its weight and provide the energy for its activities, it will eat 10,000 tons of krill in the course of its lifetime. The krill, in turn, live on the plants of the plankton, and they require 100 times their own weight. Thus a million tons of plant plankton are needed to provide for one whale.

We are told that there were once tens of thousands of blue whales in the sea; the enormous tonnages of plant plankton they must have required is almost beyond imagining. And that is by no means all. The euphausians are also eaten by other kinds of whales and by various fishes as well. Moreover, euphausians are not the most abundant among the planktonic animals. The copepods far exceed them, and they in turn must depend upon more millions upon millions of tons of planktonic plants.

This incalculable welter of activity must be powered by a comparably vast source of energy, and that, of course, is the sun as it shines its light over 143,000,000 square miles of ocean. Vanishingly small within each of the 327,000,000,000 cubic miles of ocean water are the one-celled plants that maintain their tiny, but highly complicated biochemical laboratories, adapted through eons of evolutionary development, to convert the sun's energy into the stored chemical energy that is used in turn by copepods, and herrings, and codfish, and men.

A modest man, Johan Hjort gave more than a fair share of credit to the work of others. "Thanks to the international investigations, so extensive and so excellent in quality," he wrote, "which have been carried out in recent years in all North European countries in connection with these problems, certain important general results have been established. The fluctuations from which commercial fishing populations have always suffered are due to difference in the stock of fish at various times. Only those who were engaged in the work some thirty or forty years ago can appreciate how fantastic this conclusion, now confirmed by such wide experience, seemed then. . . . The effect of man's fishing industry is undoubtedly great, but *in many cases it is much smaller than the fluctuations produced by nature.*"[4]

Johan Hjort died on August 7, 1948, at the age of seventy-nine, but not before he had been honored with doctorates from Naples, Cambridge, London, and Harvard, elected to membership in the Norwegian and French academies of science and the American Philosophical Society, and made a Fellow of the Royal Society of London. From 1938 until his death he was president of the International Council for the Study of the Sea.

In later life his interests turned to the uses of biological study as one way of serving man's need for intellectual stimulation. He concluded his lectures *The Human Value of Biology* with these refreshing thoughts:

"Those who would serve society should therefore increasingly apply the means now used for wars and class wars to the practical problem of liberating those who suffer from want of intellectual activity in their work, the greatest of all social diseases. To give man the intellectual ascendency over his tools might be one of the slogans for the advancement of this idea.

". . . Scientific methods of thought and work which once were the monopoly of a small part of society are now spreading into all kinds of human activity. Men no longer work with objects of which they know nothing. The instruments of observation and thought and the knowledge which science can offer to assist their work are placed at the disposal of workers to an ever-increasing degree. From this, freedom and happiness will follow."

---

[4] Emphasis mine (R.W.).

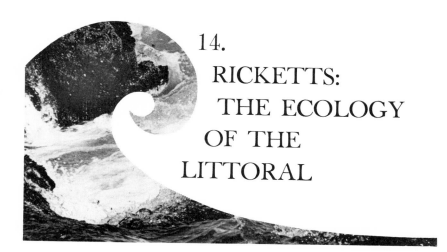

14.
RICKETTS: THE ECOLOGY OF THE LITTORAL

Just about dusk one day in October 1944, Ed Ricketts stopped work in his laboratory on Cannery Row in Monterey, California. He covered his instruments and put away his papers and filing cards. He wanted some beer, and he rolled down his sleeves before starting across the street for half a dozen bottles.

Outside he heard footsteps, and John Steinbeck shouldered his way in with a typescript in his hand. He had just finished writing a book called *Cannery Row*. Doc, the central character, was loosely patterned after Ricketts. Steinbeck wanted to know if Ed would resent any of it, and offered to make any changes that Ricketts might suggest.

"He read it through carefully, smiling," Steinbeck reported later, "and when he had finished he said, 'Let it go that way. It is written in kindness. Such a thing can't be bad.'"

One wonders if it was not Ricketts who was showing the kindness. Steinbeck's portrayal of Doc was surely one-sided. It did not touch on the most significant aspects of Ricketts' life and work, or on the fact that Ricketts was an earnest biologist who accom-

A stretch of protected outer coast near Carmel, California. One of Ed Ricketts' four major environments of the littoral.

plished—between diversions—a prodigious amount of original research.

After Ricketts died and Steinbeck, who had been a collaborator as well as a close friend, wrote the more rounded portrait "About Ed Ricketts" in *The Log from the Sea of Cortez*, he failed to mention the important fact that Ricketts, in writing *Between Pacific Tides*, had made a major contribution to marine ecology. How could Steinbeck have read through Ed's New Series Notebook No. 2[1]—which he did—without making some acknowledgment of Ed's scientific stature? Not long ago I leafed through this notebook at Hopkins Marine Station in Pacific Grove, and I was deeply impressed with the scholarship it shows. Its three hundred big ruled pages are filled with careful notes on Ed's reading— Charles Elton's *The Ecology of Animals*, for example—references

---

[1] In a holographic will dated February 28, 1940, Ed wrote in part, "To John Steinbeck. My personal notebooks, papers, and MSS in progress. . . ." After Steinbeck had cut out a few very personal entries, he deposited the notebooks in the library of Stanford University's Hopkins Marine Station in Pacific Grove. They have never been published except for a few brief excerpts that Joel Hedgpeth deciphered from the faded pencil and published in "Philosophy on Cannery Row."

to the scientific literature, addresses of specialists, and sketches and remarks about unusual marine animals.

Perhaps Steinbeck's novelist's eye was focused on the dramatic, and there was surely plenty of that in Ed Ricketts's way of life. Perhaps that novelist's eye saw only dull drudgery in the endless records of marine invertebrates that Ed compiled over the years and that were to form the basis for *Between Pacific Tides*. Perhaps "the Doc of *Cannery Row* is half Ricketts the man and half Steinbeck the author," as Joel Hedgpeth, who knew them both, has suggested.

Whatever the reason, Steinbeck did not mention Ed's book. Nor did he have much to say about Ed's habits of work. From his tales of Doc one might conclude that Ricketts' work was a sporadic thing that occasionally interrupted a procession of beer bouts, love affairs, first-aid repairs of Cannery Row characters, and psychological counseling of the girls from Monterey's most respected house.

Ed's own journals hardly support that. In his Notebook No. 2 he wrote:

"Anyone who truly has work to do will get to admiring Goethe's dictum 'Haste not, waste not.' The lucky ones who are physically and emotionally relaxed so that they can just 'be'—and enjoy life—need not have (more than enough to keep them economically adjusted and psychol[ogically] adj[usted] to a world that respects some significant outside world) but the rest of the really worthwhile people have to 'work' to build a business, a concept, or work of art, or an organization, they have to create something externally. Because I am relaxed enough to enjoy life, leisure, dawdling around talking or laughing, only once or twice a week, I must put in the rest of my life working hard and unceasingly and taking considerable exercise, if I am to avoid ill health, irritation and frustration. . . . I am usually too tense to be able to sit around enjoyably, lovingly, and listen to people talk and laugh and myself participate. I have to be up and doing. . . . So I have to put just about all my considerable energy, not in living, but in doing and in doing wisely and not frantically. I have to utilize every hour and in missionary for my way of life. Otherwise I'll get bored. 'Haste not, waste not.' "

Edward Flanders Ricketts was born in Chicago, Illinois, on May 14, 1897, and spent most of his boyhood there. At the age of six, as he once wrote to a publisher, ". . . I was ruined for any ordinary activities when an uncle who should have known better gave me some natural history curios and an old zoology textbook. Here I saw for the first time those magic and incorrect words 'coral insects.' "

When Ed was about eleven, the family spent a year in the small town of Mitchell, South Dakota, where the confluence of three creeks swells the James River on its way south to the Missouri. There in the wide-open spaces young Ed collected bird's eggs and butterflies and raised a great noisy flock of pigeons.

Ed's introduction to college-level biology came at Illinois State Normal, and from there he went on to the University of Chicago, where the only courses he took seriously were those in zoology. He stayed for two and a half years, supporting himself by tending furnaces morning and evening and clerking in a small store. He left Chicago late in 1922 without bothering to take a degree, and joined Albert E. Galigher as partner in the biological supply house that Galigher had started in Pacific Grove, California. When Galigher later moved to Berkeley with the part of the

Plankton may be easily netted from the shore. Here a biologist casts a net of fine gauze. The mouth of the net is held open by a heavy brass ring, and most of the plankton are washed down into a "bucket" clamped to the small end of the conical net. The rig is retrieved by means of a long cord attached to it.

business that provided microscope slides for student use, Ed went to the Cannery Row quarters that Steinbeck made famous as Doc's lab.

There he collected specimens of marine animals from up and down the Pacific littoral, preserving them with meticulous care and supplying them to academic biologists for use in research and teaching. That was his living. But his life was much, much more. He was driven by a kind of frenzied enthusiasm (he was said to be hyperthyroid) to know more about how the animals that overflowed the tide pools related to their physical environment and to each other. These animal aggregations were just the sort that had fired Ed's imagination as he listened to the great W. C. Allee tell of his work on animal aggregations at the University of Chicago.

Allee was a pioneer in one branch of ecology—the branch that Allee himself had chosen to call "general sociology." His concept was vast, for it extended from a consideration of the origin of life on earth to the most advanced social relations among men.

Those who had studied the social life of animals before Allee had been chiefly concerned with the social activities of single species; thus a great deal was already known about the social habits of ants, termites, and bees, in which the species had progressed to a true division of labor—as with the worker bees, the drones, and the queens.

Allee, by contrast, was interested, not in such colonies, but rather in aggregations of different species that occupied single ecological niches and seemed to do so to the mutual benefit of all. For example, Allee cited William Beebe's prospering flock of twenty-eight birds, made up of individuals of twenty-three different species, and from this he reasoned that the members of this flock must have *individual* drives toward communal life.

Allee's interest in animal aggregations began in the summer of 1915, when he and his students started making systematic collections of marine animals near Woods Hole, Massachusetts. They made careful analyses of the physical factors that seemed to influence the distribution of these animals, and eventually they saw patterns in the kinds of animals that grouped themselves together. Every summer from 1915 through 1921, Allee and his students were to be found at Woods Hole, repeating their observations and making the most careful records of them. Even today, such continued observation of one area is unusual, although with our present environmental concern it ought to be the rule.

In addition to his own Woods Hole data, Allee began to collect information about the social behavior of animals from correspondents all over the world, and gradually he was able to see a common pattern emerging. By 1930 he had written his classic progress report, entitled *Animal Aggregations,* which Ed Ricketts read and pondered over throughout his life. Allee summed up his thoughts as they stood in the summer of 1930 in these words:

"We have been concerned in this book in tracing the earliest beginnings of . . . group reactions (whether shown in overt acts or more subtly revealed), exhibited only under restricted conditions in nature which may be mimicked by properly controlled laboratory conditions. We have found that the physiology of the group considered independently from that of the individuals of which it is composed, begins simply and shows stages in development which can be arranged in various sorts of ascending series and which culminate in the group-centered division-of-labor type of society that at first glance seems impossibly remote from the life of the 'solitary' animals.

"Brilliant students of the highly social life of insects . . . have found evidence that the behavior of these societies, taken together with observations on ecological associations and the various activities that center about reproduction, indicate the existence of a fundamental tendency toward co-operation. It has been much more easy for a student [like myself] beginning with the humbler group levels to follow, from the social beginnings which he learns to recognize in almost unintegrated animal aggregations, the possibilities of the development of great social structures, and to trace their growth slowly and as yet imperfectly, but surely."

Ed Ricketts found that the coast of California was an ideal place to add new observations to the fund of information that Allee had gathered at Woods Hole. As Allee himself had said, "Along the seashore, in such favorable locations as part of the California coast, the supply of animal life is appalling. One cannot step on the rocks exposed at low tide without crushing sea urchins, sea anemones, barnacles, or mollusks."

So it was that besides making his living at collecting seashore animals, Ed Ricketts began the work that was to be published nine years later as the first edition of *Between Pacific Tides.* Joel Hedgpeth says in his "Philosophy on Cannery Row" that the book started as a sort of cottage industry. Ed's friends in Pacific Grove often went with him on his collecting trips to the tide pools, and as they began to appreciate the data that Ed was assembling on

A rich area of protected outer coast at Carmel, California. Ed Ricketts collecting.

his meticulously kept file cards, they told him that he should write up all these things in a book.

So a working plan evolved. At first the idea was for a little seashore book for beginners, and Jack Calvin, a struggling freelance writer, was to help put Ed's scientific reporting into easily understandable language and to take photographs. Ritchie Lovejoy was to make the drawings. Over the next five years, however, the book grew and grew, and as Ed read and reread Allee's papers on the physical factors that influence the marine animals of the Woods Hole region, he became more and more convinced that his own book ought to be arranged—not according to the usual biological classification of the animals into phyla—but according to the physical characteristics of the environments in which he found them.

These environments he divided into: the protected outer coast, with its rocky shores and sandy beaches; the open coast with its own rocky shores and sandy beaches; the bays and estuaries with their rocky shores, sand flats, eelgrass, and mud flats; and the wharf pilings both protected and exposed.

By the summer of 1930, the book was ready to be submitted to publishers, and in a letter to Stanford University Press, Ed argued that his approach would be novel, that it could be easily applied by the ordinary observer, and that it would have value for quick reference even for serious students and scientists. But it was nine long years before he saw the work in print.

When publication came at last, it made little difference in Ed's income. Stanford printed a mere 1,000 copies, with some misgivings about the "Annotated Systematic Index and Bibliography"—which has since proved one of the book's most valuable features. Even more discouraging, the publishers allowed it to go out of print in the years between 1942 and 1948.

But that lapse from print was by no means the greatest loss to readers of *Between Pacific Tides*. That loss occurred when the editors struck from the manuscript Ed's "Zoological Introduction"; this at the behest of one referee—a professor of zoology of narrowly pedantic turn of mind.

The professor was especially critical of Ed's discussion of an ecological observation that has since come to be known as "Gause's principle," which, thirty years after the professor's disapproval, has won recognition as an especially important ecological rule. Simply expressed, Gause's principle states that closely related organisms that have similar habits or forms are not likely to be found in the same places. In the professor's eyes, there was no way that ecologists could make anything useful out of that.

We now know that he could not have been more wrong. As late as the spring of 1973, Professor Joel W. Hedgpeth thought that suppression of Ed's discussion "may have set back marine ecology on the Pacific Coast for decades."

As it happened, Ed had not known of Gause's work, but he did know of the earlier publication of the same general idea. This had been done in 1932 by Angel Cabrera, but his paper had appeared in a little-known Argentine journal and had gone unnoticed until Ed spotted a summary of it in a 1935 issue of *Biological Abstracts*. Brief though that statement was, Ed recognized the important bearing that it had on his own studies of the Pacific littoral; and it was in the rejected draft of his "Zoological Introduction" that he suggested its potential usefulness as an aid to identifying specimens in the field.

"On the whole," he had written, "it will probably be granted by most field zoologists that competent observers might even now construct fairly accurate zonal graphs in which the tide level stations of many of the commoner forms could be plotted, based on the means of many counts and measurements over widespread areas. Obviously, we cannot make even a pretense of having done this in the tentative and approximate arrangement which follows, but which nevertheless, may prove both stimulating and suggestive.

"The limitations of any tool will become apparent sooner or later, and it may as well be emphasized at the start that it would be inadvisable for the collector to attempt any adequate identification of his catch solely by means of this classification. This despite the Cabrera 1932 law of ecological incompatibility whereby 'In the same locality . . . directly related animal forms always occupy different habitats or ecological stations. . . . Related animal forms are ecologically incompatible, and the incompatibility is the more profound the more directly they are related.' (Quoted from *Biological Abstracts*, March 1935, #4488.) But despite that probably sound generalization, ecological arrangements cannot yet, possibly cannot ever, achieve the finality so characteristic of the taxonomic order. There an animal belongs irrevocably in the one place finally assigned to it, however much that position may have been shifted about by pioneer workers. Any ecological classification will be inexact, and even its primary assignment to one may not be certain until documented by quantitative methods. However, ecological arrangements ought not to be discarded because of their inadequacies; they ought rather to be used guardedly and for what they are worth, with due regard for their limitations."

The unfortunate omission of this seminal discussion from *Between Pacific Tides* has only recently been discovered by Joel Hedgpeth, its current editor.[2] If and when a later edition appears, we should expect to see the blunder fully rectified.

From the very beginning, however, one of Ed's primary drives was to communicate, and with a published book he had achieved one way of doing that. As Joel Hedgpeth later said, "It seems to me that the most characteristic thing about Ed was his interest in communicating with people, and in his notebook I find verification of this: 'Re my concern over the deep things in sex or love or friendship or thinking or esthetics not being communicated: It gives me a feeling of waste, of futility, when I have the spark and cannot communicate it. I get frustrated and negativistic.' "

In spite of its small printing, the first edition of *Between*

---

[2] Hedgpeth made the discovery in the course of an extended study that he undertook while editing a critical selection of hitherto unpublished Ricketts papers. This volume will be published by Stanford University Press under the title *The Outer Shores*.

Pacific Tides had hardly come from the press before Ed was working on the revisions for the second, which was not published for still another nine years.

I have Ed's personal copy of the first edition on my desk. With the pages open, it still smells of formaldehyde. One page carries a long note that Ed added in neat, tiny print in India ink—new data that he had found in the literature about the copepod *Tigriopus fulvus*. The "Annotated Index" is covered with revisions in pencil, India ink, and magenta crayon. Most of these notes have been crossed out in pencil, and I would guess that Ed crossed them off as he transferred this working copy to typescript.

Stanford University Press prepared to print 3,500 copies of the second, revised edition, John Steinbeck wrote a preface for it, and Ed read page proofs and prepared the index copy. But he was not to see the book.

Between Pacific Tides went on to establish itself as an invaluable guide for tide-pool amateurs; high school teachers and university professors also found it an invaluable teaching aid. Stanford began to realize the potential of the property, and after Ed died they began looking for a literate marine biologist who might carry the work into new updated editions.

Joel W. Hedgpeth, then stationed at the Scripps Institution of Oceanography, was busy editing his own monumental *Treatise on Marine Ecology and Paleoecology* for the National Academy of Sciences, but he agreed to take over *Between Pacific Tides* as well, and he has carried it to its present classical position. Stanford now reports that, in its thirty-three-year life, the book has sold more than 50,000 copies.

Ricketts met John Steinbeck in a dentist's waiting room in 1930. Ricketts had heard that Steinbeck had taken a course in general zoology at Stanford University's Hopkins Marine Station and that he had a continuing interest in the ecological side of marine biology. For his part, Steinbeck had been "timid about meeting Ed Ricketts because he was rich people by our standards. This meant that he could depend on a hundred to a hundred and fifty dollars a month and he had an automobile. To us this was fancy, and we didn't see how anyone could go through that kind of money. But we learned."

Steinbeck's slant on marine ecology differed somewhat from that of Ricketts, for at the Hopkins Marine Station he had been fascinated by C. V. Taylor's description of the colonial life of ascidians—the strange animals that are akin both to the elemental sponges and to the higher vertebrates. The eye-opener to Steinbeck was that colonies of ascidians show an actual division of labor, with individual members of the colony performing different duties for the benefit of the colony as a whole. Thus some individuals collect and distribute food, while others protect those that do the collecting. The whole colony could therefore be viewed as a superorganism whose whole was greater than the sum of its parts. This was the basis of the "holistic" view that ran through many of Steinbeck's novels as well as his non-fiction.

In spite of these differences in emphasis, both Steinbeck and Ricketts were driven by a passion to understand the inner secrets of nature—to seek out the "deep, true things"—and many a night they spent thrashing them out with a gallon of thirty-nine-cent red wine at hand.

In addition to his revisions of *Between Pacific Tides,* Ed was also thinking about extending a similar study to the Gulf of California. Steinbeck had some money by then, and in the spring of 1940 the two laid plans for a six-week cruise to observe and collect along the littoral of the Gulf.

Steinbeck told of the cruise in the narrative portion of *Sea of Cortez* (1941), and Ricketts wrote the scientific appendix—about half the book. To verify his identifications of species, Ed went to Mexico and made intensive use of the facilities of the Universidad Nacional Autónoma de Mexico, the Institute of Biology in Chapultepec Park, and the library of the Academia Nacional de Ciencias Antonio Alzate. And of course he made full use of the facilities and worked with specialists on animal groups at Hopkins Marine Station.

The result of all this detailed work was a catalogue representing an analysis of some 550 different species of marine animals in the Gulf of California, some 10 per cent of which were new at the time of capture. This is as valuable for serious students of the marine fauna of that area as *Between Pacific Tides* is for its own province. Besides this careful detail, Ricketts and Steinbeck achieved their over-all purpose, which was to "get an understanding of the region as a whole, and to achieve a toto-picture of the animals in relation to it and to each other."

We cannot, however, leave *Sea of Cortez* without some comment on the famous but highly controversial Easter Sunday chapter—the philosophical chapter on non-teleological thinking. In essence, the argument of this chapter is that it is not helpful to look for cause-and-effect relationships in nature—or in life either, for that matter. In seeking out causal relationships there is too great a chance of falling into the *post hoc, ergo propter hoc* fallacy: the fallacy of hasty generalization that infers that because one event follows another, it must be the effect of the other.

Rather than seeking such causal relationships, they argued, it is sounder simply to accept things as they are, and having accepted them, to look for the deeper meanings that they may imply. "Understandings of this sort," they said, "can be reduced to this deep and significant summary: 'It's so because it's so. . . .' The whole is necessarily everything, the whole world of fact or fancy, body and psyche, physical fact and spiritual truth, individual and collective, life and death, macrocosm and microcosm . . . , conscious and unconscious, subject and object. The whole picture is portrayed by *is*, the deepest word of ultimate reality, not shallow or partial as reasons are, but deeper and participating, possibly encompassing the Oriental concept of *being*."

This non-teleological view, this "'is' thinking," has been criticized on various grounds, but the comment that seems to me most penetrating is that of Professor Joseph Fontenrose, a scholar of classical mythology. In *John Steinbeck: An Introduction and Interpretation*, he says that the "philosophical center" of the chapter sets up a straw man, and that it confuses final causes with efficient causes.[3]

While I cannot agree that Ricketts and Steinbeck deliberately set up any such straw man, I do think that making the distinction between final causes and efficient causes is vital. Without taking a position one way or another about how fruitful the search for final causes may be (except for certain philosophers), we can take the whole history of experimental science as proof of the value of the search for efficient causes.

A single case of an animal that Ricketts and Steinbeck themselves found in the Sea of Cortez will make the point. During World War II, American submarine strategists ran into a baffling problem, and to help solve it, they called in scientists from the Scripps Institution of Oceanography. As Helen Raitt and Beatrice Moulton report it:

---

[3] Various distinctions between final causes and efficient causes have been drawn by philosophers from Aristotle on. A good discussion of this thought is given in the article "Cause and Effect" in James Mark Baldwin, ed.: *Dictionary of Philosophy and Psychology*, Vol. I (Gloucester, Mass.: Peter Smith; 1957). In the present context a final cause implies an intent or purpose (as of some intelligence), while an efficient cause is simply "a thing which existed in time before that of which it is the cause."

"Although the scientists experimented with various types of radiation and even powerful underwater lights as possible means of submarine detection, efforts were eventually concentrated on SONAR (from SOund, NAvigation Ranging) techniques. [To "range" means to determine distance.]

"Well-known scientists were drawn from many parts of the country to conduct this research, and those with wide experience in purely theoretical physics found themselves confronted with wholly new problems as they tried to interpret the complex workings of nature. Submarine detection instruments, for example, had been plagued by a recurrent interference, a kind of rapid-fire crackling noise, under what seemed to be completely variable circumstances regarding time and location. Every conceivable cause was investigated, but the experts in underwater acoustics remained completely baffled. Finally marine zoologist Martin Johnson was called in, and he proceeded to collect various marine animals and listen to them, at the same time attempting to reconcile data on the locations of the occurrences of this interference with his own knowledge of the habits of likely animal culprits. At last he assembled enough 'snapping shrimp,' each capable of making a sharp little cracking noise by snapping shut its oversized foreclaw, to produce a facsimile of the recorded interference. Before long the picture became complete; it was found that the interference occurred only at certain depths and in proximity to the rocky bottoms where these shrimp lived. Submarine geologist Francis Shepard joined the investigations to contribute his knowledge of bottom topography, and it became possible to predict where the crackling interference would occur, and eventually to use this noise in concealing U.S. submarines. Under cover of the snapping shrimp and other predictable camouflage, U.S. submarines were able to slip undetected even into enemy harbors, and SONAR operators no longer had to worry about the mysterious noise."

Thus the search for efficient causes can be most rewarding—at least in a practical sense. Why, then, did Ricketts and Steinbeck dismiss the search for efficient causes with a wave of the hand, and philosophize at such length about final ones? We might well hazard the guess that it was largely because neither of them had had any real experience in experimental laboratory research. They were naturalists, not experimentalists. Steinbeck took one undergraduate course in general zoology. Ricketts left college after two and a half years. Original research is not normally under-

This tide pool at Pacific Grove, California, is one of many with which Ed Ricketts was familiar. On the ground both above and beneath the water's surface, note the many little black shells with white dots on them. The hermit crabs that live in these shells move these "houses" with startling speed.

taken before the graduate years, and even much of that is of a most pedestrian sort.

The Easter Sunday chapter actually had its beginnings in an essay on non-teleological thinking that Ed wrote as an undergraduate at Chicago, and we may hazard the guess that if he had gone on to do some laboratory work of his own—which would perforce have involved a search for efficient causal relations—he would have kept that undergraduate effort locked up and never permitted it in print.

But to return to animal aggregations. Those that found their way into *Between Pacific Tides* and *Sea of Cortez* were far from enough to satisfy Ed Ricketts. The Pacific stretches far to the north as well, and Ed wanted to know about the animals there too. His last notebook contains an outline of a plan for studying the west coast of Vancouver Island and the Queen Charlottes, across Hecate Strait from British Columbia. Steinbeck planned to join this expedition to the north and to collaborate in writing the book that they expected to call *The Outer Shores*. They bought tickets, Ed assembled all the necessary collecting equipment, and on the eve of their departure Ed was finishing up some last-minute odds and ends. Then . . . but let Steinbeck tell it as he did in *The Log from the Sea of Cortez*.[4]

> Just about dusk one day in April 1948 Ed Ricketts stopped work in the laboratory in Cannery Row. He covered his instruments and put away his papers and filing cards. He rolled down the sleeves of his wool shirt and put on the brown coat which was slightly small for him and frayed at the elbows.
>
> He wanted a steak for dinner and he knew just the market in New Monterey where he could get a fine one, well hung and tender.
>
> He went out into the street that is officially named Ocean Avenue and is known as Cannery Row. His old car stood at the gutter, a beat-up sedan. The car was tricky and hard to start. He needed a new one but could not afford it at the expense of other things.
>
> Ed tinkered away at the primer until the ancient rusty motor coughed and broke into a bronchial chatter which indicated that it was running. Ed meshed the jagged gears and moved away up the street.
>
> He turned up the hill where the road crosses the Southern Pacific Railway track. It was almost dark, or rather that kind of mixed light and dark which makes it very difficult to see. Just before the crossing the road takes a sharp climb. Ed shifted to second gear, the noisiest gear, to get up the hill. The sound of his motor and gears blotted out every other sound. A corrugated iron warehouse was on his left, obscuring any sight of the right of way.
>
> The Del Monte Express, the evening train from San Francisco, slipped around from behind the warehouse and crashed into the old car. The cow-catcher buckled in the side of the automobile and pushed and ground and mangled it a hundred yards up the track before the train stopped.
>
> Ed was conscious when they got him out of the car and laid him on the grass. A crowd had collected of course—

---

[4] Copyright 1941 by John Steinbeck and Edward F. Ricketts, 1951 by John Steinbeck, renewed © 1969 by John Steinbeck and Edward F. Ricketts, Jr. Reprinted by permission of Viking Press, Inc.; William Heinemann, Ltd., Publishers; and McIntosh and Otis.

people from the train and more from the little houses that hug the track.

In almost no time a doctor was there. Ed's skull had a crooked look and his eyes were crossed. There was blood around his mouth, and his body was twisted, distorted—wrong, as though seen through an untrue lens.

The doctor got down on one knee and leaned over. The ring of people was silent.

Ed asked, "How bad is it?"

"I don't know," the doctor said. "How do you feel?"

"I don't feel much of anything," Ed said.

Because the doctor knew him and knew what kind of a man he was he said, "That's shock, of course."

"Of course!" Ed said, and his eyes began to glaze.

They edged him onto a stretcher and took him to the hospital. Section hands pried his old car off the cow-catcher and pushed it aside, and the Del Monte Express moved slowly into the station at Pacific Grove, which is the end of the line.

Several doctors had come in and were phoning, wanting to help because they all loved him. The doctors knew it was very serious, so they gave him ether and opened him up to see how bad it was. When they had finished they knew it was hopeless. Ed was all messed up—spleen broken, ribs shattered, lungs punctured, concussion of the skull. It might have been better to let him go out under the ether, but the doctors could not give up, any more than could the people gathered in the waiting room of the hospital. Men who knew better began talking about miracles and how anything could happen. They reminded each other of cases of people who had got well when there was no reason to suppose they could. The surgeons cleaned Ed's insides as well as possible and closed him up. Every now and then one of the doctors would go out to the waiting room, and it was like facing a jury. There were lots of people out there, sitting waiting, and their eyes all held a stone question.

The doctors said things like, "Doing as well as can be expected" and "We won't be able to tell for some time but he seems to be making progress." They talked more than was necessary, and the people sitting there didn't talk at all. They just stared, trying to get adjusted.

The switchboard was loaded with calls from people who wanted to give blood.

The next morning Ed was conscious but very tired and groggy from ether and morphine. His eyes were washed out and he spoke with great difficulty. But he did repeat his first question.

"How bad is it?"

The doctor who was in the room caught himself just as he was going to say some soothing nonsense, remembering that Ed was his friend and that Ed loved true things and knew a lot of true things too, so the doctor said, "Very bad."

Ed didn't ask again. He hung on for a couple of days because his vitality was very great. In fact he hung on so long that some of the doctors began to believe the things they said about miracles when they knew such a chance to be nonsense. They noted a stronger heartbeat. They saw improved color in his cheeks below the bandages. Ed hung on so long that some people from the waiting room dared to go home and get some sleep.

And then, as happens so often with men of large vitality, the energy and the color and the pulse and the breathing went away silently and quickly, and he died.

Ed Ricketts was fifty-two. Had he lived his three score years and ten, he and Steinbeck would first have made their expedition to Vancouver Island and the Queen Charlottes. Perhaps they would also have made it to the Aleutians, as they had planned before the war cut that off. Both excursions would surely have resulted in notable books. Perhaps, as Joel Hedgpeth has suggested, Steinbeck might even have written the great novel of environmental concern.

What would Ed Ricketts himself have done?

I like to think that Ed, then fuller of years—perhaps with a quieter thyroid, and hopefully after doing some actual experimental work himself—would have set down his summing up in one place, for he was a communicator.

Just what toto-picture he would have arrived at, I cannot guess, except that it would have been full of deep and true things. As Jack Calvin and Joel Hedgpeth wrote in 1952, "If Ed Ricketts has achieved a trace of immortality, we believe that it is because of his ability to plant in the minds of others not facts, since many can do that, but the essential truths beyond the facts: the shadowy half-truths, the profoundly disturbing questions that thinking men must face and try to answer."

15. HARDY: FOOD CHAINS OF THE SEA

Sometimes, especially among men of great talent, we find individuals who hold to a steadfast singleness of purpose throughout long lifetimes. When those purposes are dedicated to the human good, we surely have reason to applaud. Such a man is Alister Clavering Hardy, born before the turn of the century, and now—three-quarters of the way through it—still intensely interested in the food resources of the sea. By title he is an emeritus professor at Oxford University; by interest he follows the latest advanced researches into every aspect of marine ecology and the future of the world's fisheries. Through the years his chief interest has centered on the food, the habits, and the enemies of the herring—man's most important food fish.

And while herring have been his central concern, he has been wise enough to realize that there is hardly any aspect of knowledge about the sea—its currents, its chemistry, its temperature variations, and the life that it supports, from submicroscopic plants to whales—that does not have some bearing on the success of the herring fisheries off the coasts of his native England.

As a youngster of four or five, one of Alister's greatest thrills was to be taken to Scarborough to watch the Scottish herring fleet sail into the bay to land its catch. Scarborough lies just above the 54th parallel and looks out across the North Sea toward the coast of Denmark. A third of the way across that expanse of northern ocean lies the Dogger Bank, a great shoal rising to within sixty feet of the surface of the sea, and one of the world's prime fishing grounds for those species that live near the bottom: the hake, haddock, cod, and their relatives. These bottom species are caught by trawlers, vessels that drag long baglike nets behind them over the ocean floor—much like the naturalist's dredge of Edward Forbes's day.

Closer to the British shore, where the vertically circulating currents carry mineral nutrients from the bottom, lie the herring grounds. There, near the turn of the century, one sailing lugger on a single night's drifting might take from 50 to 100 crans of herring. (One cran is 37½ imperial gallons—or from 900 to 1300 herring, depending on their size—so that 100 crans would put a boat really low in the water.) Record shots of 200 crans had been heard of, and had taken up to twelve hours just to haul aboard.

The net for herring is an entirely different affair from the dredge. It is a procession of curtainlike gill nets that hang vertically in the water, buoyed up by floats above, and held vertical by the warp—a long, continuous cable that runs below, from net to net. One net of the procession may be fifty feet from top to bottom, and something like twice as long. As the string of individual nets, strung end to end, is "shot" from the lugger about dusk, a continuous curtain is formed that may stretch out for a mile or more.

The meshes of the nets are of such size that an adult herring can push its head through, but not all of its body, and when it tries to wriggle backward, the threads of the mesh catch under its gill covers.

Long after his childhood visits to Scarborough, Hardy pictured the color of the fishing off Lowestoft. "Let us imagine for a moment," he wrote in *The Open Sea*,[1] "that we are now aboard a drifter in the autumn herring fishing, that it is two o'clock in the morning and we have just been roused up to haul in the nets. . . .

"We shall see how the fish have been caught in these very simple nets. I know of no sight in all industry to equal that of

---

[1] This and the following quotations from *The Open Sea: Its Natural History* are reprinted by permission of Houghton Mifflin Company and William Collins Sons and Company, Ltd.

the hauling of the herring nets unless it be the running of molten metal in the foundry. The ship is rolling gently; and the fishing lantern slung from the mast swings to and fro, so that all the shadows sway. One of the deck-hands winds in the warp on the . . . capstan, and the rest of us are at work on the nets, hauling them in over a roller on the side of the ship. If the catch is a good one, the nets will come up laden with fish hanging in the meshes—a mass of glistening, quivering silver. If we are lucky, too, we may see the nets as they leave the water ablaze with green fire—the phosphorescence of the sea. Once aboard each net is shaken so that the meshes are pulled open, and the herring either fly upward or fall down from the net according to the way they are facing. There is the rich smell of sodden netting; and as each net is shaken, the air is filled with fluttering silver scales, glittering in the lamp-light like a shower of tinsel. The deck is piled with fish, which from time to time are shovelled through circular openings into the holds below."

Little wonder that, as a lad of four or five, Alister was fascinated by the sailing luggers as he would watch them from the shore at dawn of a still summer morning. As the sun rose out of the sea, it flooded long miles of wavelet crescents in golden light, and row upon row of dancing golden slivers stretched away to the far horizon. In the distance the luggers looked like toys, their bare masts thin as spider webs against the pale blue sky.

Then a morning wind would spring up, smoothing long transitory paths of darkening water as the dancing slivers vanished for the moment. As the wind freshened more and more, red-brown sails slowly climbed the masts, and the sails slowly filled, and the boats turned homeward.

Sail after sail would go up, and soon hundreds of them filled the sea from the inner harbor to the horizon.

Ashore, in high excitement, young Alister would watch the preparations for the arrival of the fish. Perhaps the first boat in rode highest in the water, having had the least luck at the herring grounds. Stoically the fishermen would carry their disappointing catch, creel by creel, to the gutting box and then, shouldering their nets one by one, would lay them out on the drying green in even rectangles.

As the catches from boat after arriving boat poured into the gutting box, their blue-green backs and silvery sides shimmering in the sun, teams of Scottish girls who had followed the fleet south would descend on them with keen dextrous knives, gut them in a flash, and toss them into the waiting barrels. Then down upon the fish would come sparkling salt, showered with a flourish from a foreman's scoop. Before the day's work was done, perhaps thousands of crans of herring would have been put down in salt, the barrel heads made fast by coopers, and the whole night's catch carted off to the curers'.

Before young Alister dropped off to sleep at night, he would hear again the Scottish girls' gay laughter, and merry singing, and sly jokes. And turning toward the sea, he would think of the green watery homes where all these millions of fish had dwelled, and where millions upon millions more still dwelled, and wonder how they were born, and where they traveled, and what they ate, and who their enemies were.

In the course of his long lifetime, Alister Hardy was to discover more about these things, probably, than any other man. When in time the Scottish sailing luggers were at last gone, and had been supplanted, first by steam-, and then by Diesel-powered drifters, he would think of those old days as he relaxed before a nostalgic sketch by the late Ernest Dade; dozens of tall red-brown sails off Scarborough, lifting their slender tips to the clouds. At such a time he could hardly have guessed that his thirst for knowledge about herring would take him on a two-year voyage throughout the South Atlantic to study whales.

Born in February 1896 in Nottingham, England, Alister Hardy had his preparatory schooling at Oundle. He went to Oxford, where he earned first his M.A. and then his D.Sc. degree in zoology. During World War I he served as a captain in a Northern Cyclist Battalion and later as a camouflage officer on the staff of the XIII Army Corps. When the war ended, he returned to Oxford as Christopher Welch Biological Research Scholar, and then as Oxford Biological Scholar at Anton Dohrn's Stazione Zoologica in Naples.

In 1921 Hardy accepted a post as assistant naturalist in the Ministry of Agriculture and Fisheries, and it was there that he began his scientific studies of the herring fisheries of the North Sea. His studies could have been advanced more rapidly if a well-outfitted research vessel had been available to him, but at that time such luxury was out of the question. As the next best thing, he set about designing instruments that could be towed behind commercial vessels—steamships following normal trade routes, and the fishing vessels themselves.

These instruments had to be extremely rugged, to withstand hard usage and heavy weather, and self-contained, so to speak, so that they could be loaded and unloaded (as one would a camera) by scientific personnel ashore; thus the only responsibility that the ships' crews had was to tow them at designated locations at various depths and distances. One kind—the plankton recorder—was so successful that now, after fifty years, it is still used (with some modifications) out of Edinburgh for research purposes. We shall look at it in detail as soon as we see the kinds of information that Hardy expected to gather with it.

As luck would have it, just as he was getting a good start on this project, Hardy received an offer that no zoologist with an interest in the sea could refuse. He became chief zoologist aboard the Royal Research Ship *Discovery*, and during the years 1925 to 1927 he worked aboard her while she plied her way back and forth across the South Atlantic between Cape Horn and the Cape of Good Hope, and down to the Weddell Sea with its awesome bergs and pack ice.

The principal objective of this *Discovery* expedition was to study the life of the whales of the South Atlantic, the plankton on which they feed, and the many other characteristics of the waters of this distant sea. In all, more than 12,000 samples of plankton were taken and later analyzed in minute detail.

Like the *Challenger,* the *Discovery* brought back so much information in the form of records and specimens that it took numerous specialists many years to complete all of the reports of the venture. Indeed, although the *Discovery* returned to England in 1927, it was not until 1966 that the thirty-fifth and last volume of the *Discovery* Reports was finished.

It was most unfortunate for public understanding that a popular account of this great adventure was delayed even longer. Not until 1967, in fact, did a narrative account of this expedition of 1925–7 appear.

But it was well worth waiting for. Hardy, by then Sir Alister Hardy, honorary fellow of Exeter and Merton Colleges, Oxford, published *Great Waters: A Voyage of Natural History to Study Whales, Plankton, and the Waters of the Southern Ocean.* He had waited forty years to write the book for two reasons: he wanted the official reports to be complete and in print, and he wanted to be able to relate the findings of forty years earlier to those of later voyages and more recent researches. The result is a marvelous book for the general reader, packed with fascinating observations about this little-known southern sea.

What, then, have the life histories of whales, wandering among the Antarctic ice floes of the Weddell Sea, to do with the herring fisheries off the coast of England? Part of the answer lies in the universality of the world's oceans: the major influences of the environment upon a whale in the south are not radically different from those influences upon a herring in the north. More specifically, the diets of whales and of herrings are surprisingly similar. We know, of course, that most species of whales do not eat large fishes: rather, they live on krill, small planktonic animals that they filter out of the sea water in enormous volumes with the meshes of their baleen plates. It was during the *Discovery* expedition into the South Atlantic that Alister Hardy found that the krill in that region consists almost exclusively of a single species, the little shrimplike *Euphausia superba*.

The principal food of the herring of the North Atlantic is also a shrimplike animal, in this case the copepod *Calanus finmarchicus*—a rather transparent little organism about the size of a grain of rice. When the food of the *Calanus*—the microscopic diatoms and flagellates of the North Sea plant population—is plentiful, the *Calanus* in turn become so abundant as to tinge the sea with a reddish blush for miles around. Thus, to begin his attack on the problem of the herring fisheries, Hardy had to learn as much as he could about *Calanus*. This was like opening Pandora's box. Each new discovery was a steppingstone to a dozen others, and new knowledge branched and proliferated like the nine-headed hydra of Hercules.

In his reach for a true grasp of the mystery of *Calanus*, Hardy was led to a study of the whole free-floating population of the world's oceans.

After a great many years of his own original investigations, and after close study of the investigations of hundreds of other naturalists throughout the world, Sir Alister Hardy condensed the essence of these findings into one great book. He called it *The Open Sea: Its Natural History*. It is beautifully illustrated with a profusion of line drawings and water-color sketches that Sir Alister made from living specimens at sea in the course of his many voyages. (He is still, in fact, an ardent water-colorist.) Of *The Open Sea*, Joel Hedgpeth has said, "This is one of the great books about the sea written in our generation. Although British in orientation, it is the best layman's introduction to marine biology that we have."

As to the word "plankton" itself, Sir Alister has called it

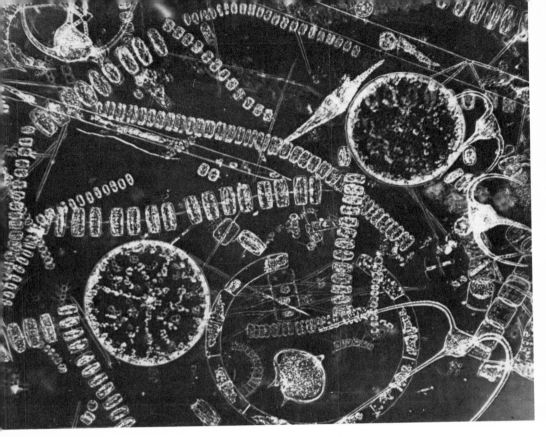

These tiny marine plants are food for the marine animals. They are here shown magnified 160 times.

"one of the most expressive technical terms used in science and is taken directly from the Greek. . . . It is often translated as if it meant just 'wandering,' but really the Greek is more subtle than this and tells us in one word what we in English have to say in several; it has a distinctively passive sense meaning *'that which is made to wander or drift,'* i.e. drifting beyond its own control—unable to stop if it wanted to." This term applies to all the plants and animals that drift freely in the ocean.

The term "plankton" was not coined until toward the close of the nineteenth century—by Viktor Hensen in 1887, as we have seen. The biologists of the middle of that century were willing to go along with Ernst Haeckel, the famous German zoologist, in dividing all plants and animals of the ocean into two great groups: those that could swim about freely (the nekton) and those that live at the bottom of the sea (the benthos).

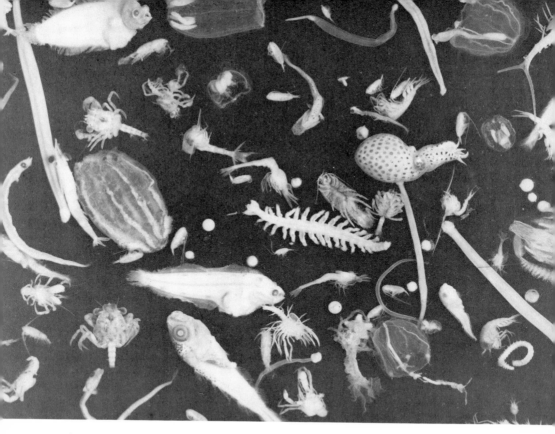

These small animals are food for the fish of the North Atlantic. Shown here at approximately 75 per cent of natural size, they are about 100 times larger than the microscopic plants on which they feed.

It was in 1828, or more than a half-century before the term "plankton" was coined, that a member of the group was first identified. In that year J. Vaughan Thompson, a British Army surgeon and amateur naturalist, invented a novel tow net which continues to be useful (in somewhat improved form) to the present day. Thompson put together a small cone-shaped net of fine gauze, held open at the mouth by a circle of stiff wire, and closed at the rear by a jar, which has since come to be known as the bucket. When Thompson towed his net behind a boat off the shores of Ireland, water flowed through the mouth of the net and out through its fine meshes, and many tiny plants and animals were caught on the gauze and washed down into the bucket. When he hauled in his net at the end of a run, Thompson's bucket was alive with the kinds of plants and animals that drifted about near the surface of the sea—the plankton.

Especially interesting among the animals that Thompson collected were some little bristly creatures that he first took to be adult animals, but which he was later able to show were the larval stages of barnacles and crabs. And he demonstrated that the barnacle, even with its adult sessile habit and limestone shell, was not a mollusk at all, but rather a crustacean.

Today Thompson's few original writings are all but inaccessible, for they appeared in the most limited printings and are now closely guarded in a few libraries. Still, his work is of the highest quality; Charles Singer has written: "It has been said of him that 'no great naturalist has written so little and that so good.'"

After Thompson, the scientists aboard the *Challenger* and Hardy aboard *Discovery* (and in later years *Discovery II*) added greatly to our knowledge of the enormously complex and abundant life of the plankton.

First, as we have already seen, the microscopic green plants of the sea (chiefly the diatoms and the flagellates) form the source of energy for all other life in the sea and, for that matter, a significant part of the animal life on land. Through photosynthesis they convert the energy of the sunlight into sugars and starches that can in turn be converted by animals into energy of motion or stored as fats for use at a later time.

The scale on which this takes place is really beyond comprehension. We can give numbers, but it is next to impossible to grasp them. What does it mean when we are told that each year the oceans of the world produce something like 150,000 million tons of these tiny green plants? Still, it is a useful figure, for it gives ecologists a starting point for at least a rough estimate of the ocean's potential yield. But yield of what? It depends on how high on the food chain our interest focuses.

For each step in the food chain, roughly 90 per cent of the original mass is lost. For example, the principal food of adult herring is *Calanus*, which graze off plant diatoms. Thus for each thousand pounds of diatoms, we can expect a hundred pounds of *Calanus*, and from them, ten pounds of herring. However, if we should prefer codfish, which eat herring, then our ten pounds of herring would produce only one pound of cod. So it takes half a ton of diatoms to yield a pound of cod.

That sequence, however, is only one thread of the food chain of the herring, and to consider it alone is a great oversimplification. The actual situation is more properly called a food *web,* which is made up of many, many such chains. Thus a herring will eat many different kinds of things in the course of its lifetime, which may last for seven or eight years in British waters and considerably longer than that off Scandinavia. As a tiny fingerling, a herring will eat a variety of things, such as the larvae of mollusks, *tintinopsis, peridinium, pseudocalanas,* and even some diatoms. After it has grown a bit larger, it turns mostly to *pseudocalanus.* At a length of four inches or so, it likes decapod larvae, *acartia,* and little sand eels. The adult herring greatly prefers *Calanus,* but will also eat half a dozen other animals, including even a species of modified sea snail.

But having eaten all these various creatures, the herring itself is the prey of voracious larger creatures. Besides man and his drift nets, the principal predators are codfish and porpoises, and even some kinds of whales, as well as sea gulls and gannets, diving with their half-closed wings to swallow herring whole.

If this web within which the herring preys and is preyed upon seems complex, how vastly more complex is the entire food web of the world's oceans! We usually think of the creatures of the plankton as being very small, and this is true of most of them. Many of the minute plankton plants measure only about one

This sketch shows the principal feeding relationships between herring of different ages and the members of the plant and animal plankton on which they depend. The arrows indicate who eats whom.

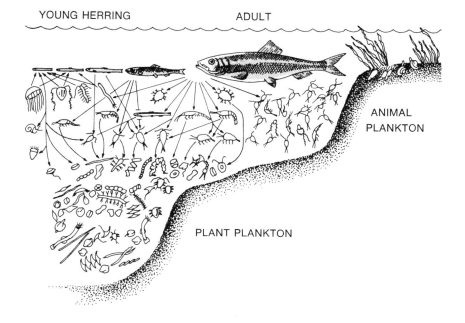

ten-thousandth of an inch, and some of the little plankton animals that graze on them are barely visible. But properly included among this drifting population are the largest jellyfish, which may be a yard or more across.

It is indeed a catchall classification, for included in it are floating fish eggs, young fish that are carried about more by the currents than by their own powers, the larval forms of mollusks and barnacles, together with those of crabs and many worms. There are little jellyfish of many shapes and colors, basically transparent, but often with tinges of green or orange or violet or red or a dull white; fringed with long cilia or short, straight or wavy, and pointing either up or down depending on the species. There are syphonophores and comb jellies and sea butterflies, and all manner of copepods from the limit of visibility to those an inch or more in length.

There are shrimplike creatures of mottled brown and red and blue, a great many of which can be seen in clear detail with a simple hand lens, and larval stages of crustaceans with suggestions of the ten legs they will have as adults; and half-red and all-red prawns; and squid that can change from one brilliant color to another. Not to mention the giant squid. One of these, found stranded at the Bay of Nigg, Aberdeen, was a dozen feet long. Thus not all plankton are small, by any means.

Every one of these fantastic creatures depends ultimately on the tiny plant life of the plankton. These little plants, in turn, depend for their existence on the inorganic nutrients in the water. Only through the mediation of these essential elements can they use the energy of sunlight to build carbohydrates from carbon dioxide and water. Their requirements do not differ greatly from those of the important land crop plants, and just as a farmer's corn yields better when adequately supplied with nitrate, phosphate, and sometimes added potassium, so too do the plants of the sea flourish best when these essential nutrients are abundant. Some of these nutrients are discharged from estuaries into the sea, but a critical proportion of them come from the decomposition of waste organic matter that has sunk to the ocean floor.

Moreover, since the marine plants can survive *only as far below the surface as sunlight can penetrate,* there has to be a mechanism for moving the inorganic nutrients from the depths to the sunlit zone. Again, the mechanism is complicated in the

Sometimes predators obtain their prey in unusual ways.

extreme, being influenced by temperature differences in the water, by the great currents such as the Labrador Current and the Gulf Stream, and by the flow, known as upwelling, that takes place along the continental shelves. Of these, upwelling is locally the most significant, especially along western shores. But in spite of all the complications, it is entirely clear that marine plants are most plentiful when the inorganic nutrients are most abundant, and in turn that small marine animals are also most plentiful when the little plants "bloom."

It was these basic facts that led Alister Hardy as a young zoologist working for the Ministry of Agriculture and Fisheries, and later when he was called to the chair of Zoology and Oceanography at University College, Hull, to develop his two famous instruments for studying plankton populations.

The first of these was known as the Hardy plankton indicator. In principle it was like J. Vaughan Thompson's plankton net of a century earlier, but it was far more rugged. It consisted of a cylindrical metal tube about a foot and a half long, shaped like a small torpedo, and equipped with a yoke for attaching a tow line. Below the nose of the tube a diving plane was fixed, resembling the pontoons of a seaplane, so that no additional weights were necessary. At the rear of the tube three stabilizing planes were mounted symmetrically, like the fins of an aerial bomb. The front end of the tube had a constricted mouth, and the net proper was simply a circular disk of sixty-mesh silk gauze mounted on a ring. After a tow, the gauze circle could be quickly removed for inspection and replaced by a fresh one as needed.

Hardy's first use of the plankton indicator was as a search device to guide herring fishermen to the most productive shoals. He reasoned that an abundance of herring food should mean an abundance of herring—although to begin with he was not certain which element of the herring food chain would be the most useful indicator. His first trials were therefore aimed at learning the relation between the kinds of plankton most abundant and the size of the herring catches. He was also intrigued by the fishermen's lore that "weedy water" should be avoided because few herring were likely to be found in it.

In *The Open Sea,* Hardy described how these first trials were carried out. "Fishermen were asked," he wrote, "to use it to sample the sea where they fished, and then to wrap up the plankton-coated disc in a piece of calico provided and drop it into a tin of preservative fluid (formalin). Later, when they had hauled in their nets, they filled in a printed label giving the date, position, and number of herring caught. The idea was to get a series of plankton samples together with records of the catch of fish."

The first fourteen records that were obtained presented a curious anomaly. The gauze disks from seven of these trials were distinctly green, indicating an abundance of plant plankton, which in turn would seem to suggest an abundance of the favorite herring food *Calanus*. The herring catches corresponding to these seven green disks averaged only 3 crans. In the other seven trials the disks were blank or colorless: the plant plankton were very scarce there. But the corresponding herring catches averaged 25 crans—more than eight times greater!

The "weedy water" lore surely seemed to be confirmed. But why? Hardy offered two possible explanations. First, it seemed reasonable that in those areas where the green plant plankton were abundant, there were simply not enough *Calanus* present to keep them grazed down; and by the same token not enough *Calanus* to attract many herring. Moreover, it seemed likely that time lags would occur; it might take some time for abundant plants to attract abundant *Calanus,* and some time more for abundant *Calanus* to attract abundant herring.

Reasonable as this seemed, Hardy was far from satisfied with it, and he set about making direct correlations of the numbers of *Calanus* caught on the gauze with the catches of herring in the same waters. In eleven pairs of observations that he made in 1932, he found that when the *Calanus* count averaged 53, the herring catch was $14\frac{1}{2}$ crans. In contrast, when the *Calanus* count averaged 353, the catches averaged $31\frac{1}{2}$ crans.

These are only two examples of a great many useful correlations that Hardy's plankton indicator made possible. As a fish-finder, however, it had the theoretical disadvantage that it could not detect the herring themselves, but only the elements of the plankton population that are usually associated with them. When the echo sounder (analogous to SONAR) was developed, it became immediately evident that its sound reflections could provide direct indications of the density of herring shoals. Practical tests soon showed the superiority of the echo sounder for fishermen's use. But the plankton indicator is still used by a number of research laboratories for their continuing surveys of plankton populations.

Another ingenious Hardy invention is the continuous plankton recorder. Like the indicator, it is a kind of net that can be

A wealth of information about the food of North Atlantic herring has been gathered by towing this Hardy continuous plankton recorder behind commercial ships in passage.

towed, usually by commercial ships, but unlike the indicator, it continuously samples the plankton for mile after mile along the ship's route. As it is towed through the water, an external propeller drives a continuous ribbon of silk gauze across the path of inflowing water. An opposing roll of fresh silk gauze unwinds at the same rate, and the two strips are pressed together to imprison the plankton as between the layers of a sandwich. The composite strip is finally wound onto a roller immersed in a tank of formalin preservative. The speed at which the silks travel can be varied from one inch per half mile of towing to an inch per five miles. The mouth of the sampler has an area of three-quarters of an inch, so that in a mile of towing the instrument filters some 150 gallons of sea water. It is a rugged device, weighing more than 150 pounds.

Hardy first used it aboard R.R.S. *Discovery* in the years from 1925 to 1927, and the Edinburgh Oceanographic Laboratory of the Scottish Marine Biological Association has used it for many years over many thousands of miles between the coasts of Scotland and those of Denmark, Norway, Iceland, and Newfoundland. For more than four decades it has been an outstanding aid to furthering our understanding of the ecology of the plankton, especially as it relates to the herring fisheries.

A critical question remains. What are the prospects for the future of the herring fisheries? We readily see how bafflingly complicated the ecology of the herring is, and how difficult it is to understand the interlocking influences upon the present, to say nothing of predicting the future.

In these days of concern about man's influence on the environment, it has become the fad to seize upon some one aspect of man's behavior and argue that *that* is what threatens the world's fisheries—or whatever. Thus the cries that pollution of the streams and estuaries that flow into the sea threaten all life in it. Or that overfishing will lead to a decline and the eventual demise of the fish populations. Or that efforts at conservation are futile anyway, because the human race is exploding at such a rate that nothing but starvation lies ahead.

There is some basis of fact at the root of all these arguments, but the gravest threat to solving the real problems that the real ocean world presents lies in becoming hysterical about them. Surely this cannot but lead to an attitude of defeatism, diverting our attention from truly significant solutions—difficult though they may be.

Let us say that the herring yield off Scotland's Firth of Moray has declined—as it certainly has—since the lush days when the "creels of silver herring turned into creels of silver crowns." Was the decline the direct result of overfishing? This seems highly doubtful. More probably it was due to many causes: changes in the direction of major ocean currents; changes in concentrations of inorganic nutrients brought to the surface by upwelling; increased predation upon adult herring by cod, gulls, and gannets; decreased diatom populations followed by smaller *Calanus* populations; more devastating attacks on herring eggs and fingerlings by jellyfish, sea gooseberries, and arrow worms; changing ocean currents carrying fingerlings to inhospitable locations; unfavorable changes in the temperature of the ocean water itself; and so on and on and on.

The essential point is that blaming overfishing for the declining fish population is not merely a hasty oversimplification; it is a most serious threat to our motivation to learn what the real reasons for such a decline may be and to seek workable solutions to the problem.

A modern development of the Hardy continuous plankton recorder. Known as the Longhurst-Hardy recorder, it is used to study the vertical distribution of plankton. It is here shown aboard the David Starr Jordan, *a high-seas research vessel operated by the Fishery-Oceanography Center at La Jolla, California.*

In the edition of *The Open Sea* that was published as recently as 1965, Sir Alister Hardy put the complexity of the problem in cogent terms:

"My concluding point unfortunately strikes a note of alarm. The last two autumn seasons 1955 and 1956 have seen an appalling decline in the landings at Yarmouth and Lowestoft. As Dr. Hodgson says, writing . . . in *The New Scientist* (1956): 'The whole fate of the East Anglican fishery, once the world's greatest herring harvest, is at stake.' He believes that this shortage of fish can be traced to the terrific new onslaught being made in the stock by trawlers, both for the young fish east of the Dogger, and for the adults at the eastern end of the Channel in December. Other leading experts, however, are not at present agreed as to the cause. Some suggest that the massacre of very young herring east of the Dogger cannot be the cause of our shortage because, they say, it is our older fish that are failing to turn up and not the young, which are still coming in reasonable numbers; others point out that the heavy trawling in the Channel is actually taking no more fish from the stock than used to be taken in the East Anglican drift-net fishery in the old days when some two thousand vessels took part. Further there are certain changes in the stock which lead some to believe there are at work natural causes which are more powerful than man's efforts. It is unfortunate that we must leave the problem in such an unsatisfactory position as to its solution."

Since Sir Alister wrote, however, there have been new developments that suggest that problems of even this great complexity may yet be solved. During World War II a new concept arose which led to attack by teams made up of scientists from many different fields upon a given problem. This strategy was outstandingly successful during the war, it has been under continuous development ever since, and it is now coming into use in studying tomorrow's problems of our ocean resources. Soon we can expect to see, probably through international cooperation, the careful and considered recommendations representing the best thinking of teams of biologists, chemists, physicists, engineers, statisticians, politicians, and all the related vocations and trades. "The sea . . . ," as James Fraser wrote in *Nature Adrift*, "will give better and better yields as we learn more and more about it and learn to exploit it in a controlled and rational way."

16.
ZOBELL: MICROBES OF THE SEA

Among the dramas that biology unfolded late in the nineteenth century, the story of microbes is one of the most fascinating. In the 1870's and 1880's the great Louis Pasteur proved that spoilage of wine and of beer wort was caused by contamination by wild microorganisms that entered from the air; he also demonstrated that these fetid organisms could be completely killed by gentle heating. Overthrowing the long-held dogma of the day, he showed that these microbes were never generated spontaneously, but always arose from preceding living ancestors.

Spoiled wine and spoiled beer were considered to be "diseases," and it was only natural that Pasteur should next look for other microbes that might cause other diseases. He found that silkworm disease was caused by a microbe, as were anthrax and chicken cholera. In addition to discovering these microbes, he devised inoculation techniques that blunted their attack. Pasteur's magnificent breakthrough was seized upon by other alert minds: Robert Koch, Ferdinand Cohn, and Joseph Lister made great strides in developing and expanding this new science of bacteriology. After them came Roux and Behring, Metchnikoff, Theobald Smith, and Bruce and Ross and Grassi, Walter Reed, and Paul Ehrlich; and their triumphant fights against Texas fever, sleeping sickness, malaria, gonorrhea and syphilis, and many more.

At about the same time, bacteriologists also learned that dead organic matter was broken down into its inorganic constituents by bacteria, but that was taken much as a matter of course; its vital importance was not yet clear. Nevertheless, it went on continuously everywhere in nature; and if these beneficial decomposing bacteria were not constantly at work, life on earth would slowly expire. Eventually all the available carbon and oxygen and nitrogen and hydrogen atoms on earth would be tied up in the bodies of dead plants and animals. No free elements would remain from which new protoplasm could be built. A desolate earth would be covered with dead plants and animals from which no new life could arise.

Of course, as the science of bacteriology matured, more attention came to be paid to the beneficial microbes, and it was eventually recognized that the beneficial ones were far more numerous than those that cause diseases (the pathogens). Nevertheless, the main thrust of bacteriology continued to center on medical applications.

It is not surprising, then, that men did not hurry to find out whether bacteria occurred in the oceans, and such early attempts as were made were mostly carried out by ship's doctors as a pastime. To break the monotony of long voyages, they would throw tethered bottles over the rail, retrieve sea-water samples from alongside, and look for microbes in them by the standard methods of medical bacteriology which they knew so well.

They inoculated the usual agar plates (prepared with nutrient media in fresh water), and incubated them at the body temperature of 98 degrees Fahrenheit—the temperature at which the pathogens grow best. And they saw that colonies of bacteria grew lustily. The sea water was alive with microbes! But what kind? Further examination showed that they were chiefly *Escherichia coli,* a normal inhabitant of the human colon that is usually quite harmless. So there were indeed microbes in the sea—microbes that had been deposited near the ship by the ship's own crew.

We can only be amazed at the crudity of these early attempts.

Who in his right mind would try to culture *marine* bacteria at 98 degrees, knowing that the average temperature of the ocean is more like 40 degrees? Or would try to culture marine bacteria in a nutrient medium based on fresh water, knowing that the salinity of sea water is some 3.5 per cent?

Obviously such early fumblings led to grossly misleading results, but it is still somewhat surprising that it was not until the 1930's that marine bacteriology took the first steps toward becoming a mature science. But that long delay was not altogether due to fumblings like those of the ship's doctors. Much of it was due to the extraordinary difficulties that beset such study.

In the first place, if you do find a microbe in the ocean, how do you know that it is a true marine microbe, having its normal home in the sea, and not one that has been discharged from the mouth of a river, or dropped by a bird, or carried by rain borne on offshore winds? To be certain, you must take your samples from far out at sea, and that requires a full-scale expedition.

You must also have facilities for examining your sea water samples on the spot. You cannot preserve them in formalin, as Sir Alister Hardy did with his plant and animal plankton, and study them at leisure, for the mere shapes of your microbes do not tell you very much about them. True, you can note that under a magnification of 1000× they look like little spheres, or rods, or coils. And with the necessary equipment you can tell that your rods, for example, are about four ten-thousandths of an inch long and four hundred-thousandths of an inch wide, but that is about all. To really identify your little creatures, you must make complicated tests of their physiological functions. And that is virtually impossible unless you have laboratory facilities aboard your research vessel.

Small wonder, then, that marine microbiology did not begin to come into its own for half a century after Pasteur's pioneering work. Small wonder, too, that only a few hardy souls ventured into a field so fraught with difficulties.

To see something of these difficulties and how they have been overcome, let us take an informal glance at the career of an outstanding leader in the field—long since recognized as a world authority—Dr. Claude E. ZoBell of the Scripps Institution of Oceanography of the University of California.

Claude ZoBell was born in Provo, Utah, in August 1904, and spent his boyhood on a farm east of Rigby, Idaho, some fifty

Dr. Claude E. ZoBell, professor emeritus of marine microbiology at Scripps Institution of Oceanography, prepares a nutrient-agar medium for culturing marine microbes.

miles west of Grand Teton National Park. After graduating from Rigby High School, he earned his B.S. and M.S. degrees from Utah State University and his Ph.D. in bacteriology at the University of California in Berkeley. He had always been strongly attracted to teaching, and planned to make his career at Berkeley in the well-established field of medical bacteriology. His faculty advisers were all in favor of this, but suggested that he would do well to broaden his background by spending one postdoctoral year at Scripps, studying the then new marine bacteriology. He was quick to agree, and soon began work in a small laboratory in La Jolla, close to the rolling Pacific where he could often hear the waves beating against the sand.

The one year stretched into many, for Claude ZoBell became so interested in the numerous problems that the microbes of the sea confronted him with that he continued at La Jolla for more than forty years. He rose rapidly through the academic ranks, and when I saw him late in 1972, he told me with a twinkle in his eye—and with strong emphasis on the key word—that he had recently been *promoted* to Professor Emeritus.

One of the first problems that he faced was how to take representative samples of sea water at various depths with the least possibility of contamination from other sources. Various kinds of water samplers had been devised by Russell, Issatchenko, Wilson, and Gee, but these were all subject to some chance of contamination, and altogether ZoBell spent ten years experimenting with these and various modifications of them before he developed what he calls the "J-Z sampler."

One version of this ingenious device has as its receptacle a stout citrate of magnesia bottle, which is closed at the mouth by a rubber stopper. A hole is bored through the center of this stopper, and an L-shaped glass tube is tightly fitted into the hole. We may visualize a citrate bottle with a rubber stopper carrying a glass tube that emerges vertically from it and then bends gently to point horizontally to, say, the right. A rubber tube is now placed over the end of the glass tube like a sleeve, and into the other end of the rubber tube, a second glass tube is inserted. Toward the outer end of the second glass tube a constriction is drawn in a flame, so that it may later be hermetically sealed off.

Next, the entire assembly is sterilized under steam pressure in an autoclave, and the constriction in the outer glass tube is heated and sealed off. Finally a file mark is scratched on this outer glass tube, so that it will break at the mark when smartly rapped. Any number of these J-Z samplers may be made up in this way, ready to go to sea. Of course, when the bottles cool after they have been taken from the autoclave, a partial vacuum is formed inside, and sea water will be sucked in when the end of the glass tube is broken off.

In use, the J-Z samplers are cradled in a metal frame which can be connected to a standard hydrographic wire for lowering into the sea. The key feature of the whole assembly is that in use the rubber tube is bent into a U shape and kept in that position by clamping the outer glass tube to the frame. The frame is provided with a lever at one side, so that a messenger weight sent down the hydrographic wire will break the glass tube at the file mark. When this occurs, the U-shaped rubber tube flips straight like a bent piece of spring steel, and the intake end of the sampler is now several inches away from any contamination that might come from the carrier or the wire.

So much for the operation of a single sampler in its carrier frame. Naturally, it takes time and costs money to lower a hydrographic wire, especially in deep soundings, and ZoBell further devised a gang carrier, any number of which can be mounted in tandem on a wire as it is lowered. When such a procession has been lowered and everything is ready, all that needs to be done is to drop one messenger weight down the wire from the top. The weight strikes and activates the first sampler by breaking the glass tube at the file mark. Immediately a second messenger mounted at the bottom of the first carrier is released, and that drops and activates the second carrier, and so on down the whole length of wire. Each time the taut wire would twang, ZoBell would have a new sample, and in a short time he could collect samples from, say, 10, 20, 50, 100, and 250 fathoms down. But when the last sample bottle was recovered, ZoBell's task would be just beginning. If the ship's facilities permitted, samples could for a short time be preserved in cold storage. Otherwise, all of the planned laboratory examinations would have to be carried out immediately. ZoBell had long since found that marine bacteria can die overnight and that results from unfresh samples were often highly misleading.

The sea water itself, however, is not the preferred location for the microbes of the sea. Most of them are found in the mud of the ocean floor, and it is therefore most important to be able to take mud samples; preferably these are taken as long cores, so as to reveal how deep into the bed the bacterial action extends. Various mud samplers have been devised, among which the Emery-Deitz rig deserves special mention. In essence it is simply a hollow steel pipe with a sharpened nose, and is weighted with three hundred to six hundred pounds of lead. It is lowered by a wire cable paid off from a ship's winch, and when it has descended to about three hundred feet from the bottom, it is allowed to fall very rapidly until it strikes the bottom. With it, core samples fifteen feet long have been successfully taken.

So that slices at various depths can be isolated for examination, the steel pipe is provided with a celluloid liner that can be readily slipped out. The liner is not a closed cylinder, however; rather, it is formed by rolling a rectangular strip of celluloid into the shape of a cylinder, which is then inserted into the pipe. After a core has been taken, the celluloid liner is slipped out of the pipe, the celluloid straightens itself out, and the cylindrical core lies bare for slicing into disks.

In a very great many tests using water samplers and mud samplers such as these in many different areas of the world, Zo-

Bell and others have found that the bottom muds contain hundreds of thousands of times more bacteria per unit volume than the waters that lie above.

The reason for this did not remain a secret for very long. Over and over it was observed that marine bacteria have a strong tendency to stick to solid surfaces. Moreover, throughout the ocean there is a steady rain of particles denser than sea water that sink slowly to the bottom: billions of silica shells of dead diatoms, horny shells of crustaceans of all sorts, insects and other land animals that fall into the sea, and so on and on. It turns out that the highest concentration of bacteria to be found anywhere in the sea is right at the interface between the water and the mud, and that is the site of the most vigorous biological activity.

Meanwhile, on the ship above, ZoBell prepares to make his microscopic and physiological examinations of his mud and seawater samples. This, as we have seen, has to be done promptly. It is easy and convenient aboard a well-equipped research vessel in a calm sea, but the sea is not always that cooperative. The work must go forward irrespective of the weather, and on many occasions ZoBell has had to carry on his work while his ship was being heavily buffeted by winds and waves. Let us watch him as he performs high-precision laboratory work while he and his equipment are vibrating, pitching, and rolling in a heavy sea.

He has been allotted only a very small space aboard the vessel—some ten square feet of laboratory room, and perhaps twenty cubic feet of storage space, for all his samples and equipment. Therefore the uses of the area have been most carefully planned. He will do his work seated on a stool that has been firmly fixed in place, and he will hold fast to it as the ship rolls, either by clamping his legs around it or by strapping himself down.

Every item of equipment he will need is within reach from his position on the stool. There has not been room for an autoclave, so he has already sterilized those items that require it in a pressure cooker in the ship's galley. The sterile pipettes, and test tubes, and culture bottles (flat-sided prescription bottles that have already been charged with nutrient bacto-agar) stand arrayed before him in special flat-bottomed holders. Each reagent bottle sits in a hole that has been bored to the right size in a fixed block of wood. Likewise microscope slides and cover glasses. The microscope itself has been clamped down to the benchtop. An alcohol burner for flaming the mouths of culture receptacles, and for sterilizing needles, has already been lit, and its blue flame wavers uncertainly as the ship rolls.

For every sample of sea water, or of mud from the ocean floor, ZoBell will have two essential tasks. The first will be to estimate the total number of microbes per unit volume of the sample, and the second will be to start cultures from which colonies of pure descendants of each kind of microbe may later be grown.

He will make his counts under the microscope. Strapped down to his stool, his upper body swaying fluidly in response to the ship's motion, he reaches for the first sample with one hand and for a small pipette with the other. He draws a few drops into the tip of the pipette by mouth suction, taking care that no sudden movement raps the pipette against his teeth. Closing the top of the pipette with his forefinger, he puts down the sample bottle and picks up a microscope slide. When he has transferred a few drops to the slide, he drops a cover glass upon it and places the slide on the microscope stage.

Now he is ready to count his microbes. His field of view has been marked off in squares of known area, and ZoBell will simply count the organisms that appear in several of these little squares. Simple multiplication will give him a good approximation of how many microbes are in the entire field. As he puts his eye to the ocular he braces himself firmly, for it would be awkward to jam his eye against the eyepiece. With the microscope in focus he begins his count of the first square. If a sudden lurch of the ship causes him to lose his place, he must begin again, for he is intent on accurate observation.

His count of gross numbers of bacteria now completed, he turns to his culture media, in which he will grow new colonies. He unscrews the cap of a flat-sided prescription bottle that has already been charged with nutrient agar, flames the mouth of the bottle to sterilize it, transfers a sample by pipette to the nutrient, and closes the bottle. If these operations require good coordination even on the steady floor of a shoreside laboratory, they are just so much more trying aboard a rolling ship.

The research plan has called for the inoculation of many culture flasks, for ZoBell wants to be able to isolate and identify as many different species of microbes as he can. He will also make parallel runs using different culture media, since some species grow well in one and not at all in another. At the least he will want two media: his own standard containing sea water, bacto-peptone, ferric phosphate, and bacto-agar; and another that may perhaps contain some glucose. But whatever the routine may be for any particular day, the wear and tear on human nerve and muscle will have been severe.

ZoBell has been the center of many such scenes, on many research vessels, at many different times, in many different areas of the world's oceans. He has made cruises aboard the *E. W. Scripps* to study marine microbes in the Gulf of California; was a member of the Danish *Galathea* round-the-world deep-sea expedition; cruised aboard the New Zealand Navy vessel *Tui* around South Island and Milford Sound; and was a member of the *Stranger* expedition to the South China Sea, the *Doda* cruise to the Marianas Trench and the Philippines, and several *Argo* expeditions, including Depac X to the Japan Trench.

Ashore, he has made a special effort to assemble and digest the publications of marine microbiologists throughout the world, and his spacious office at La Jolla is lined with bookshelves to the ceiling. His own scientific publications number more than two hundred, and he has distilled the essence of this work—over more than forty years and throughout the oceans of the world—into two books. The first of these, *Marine Microbiology,* was published in 1946. In the fall of 1972 he told me that it was then sadly out of date; he is at work on a new book that will bring the subject into today's focus, but, because of his many other commitments, the project is going more slowly than he would like.

He was kind enough to give me a capsule summary of some of the more important conclusions that marine microbiologists have so far arrived at, and I list them below, in no particular order.

- In the upper layers of the ocean, where the photosynthetic process takes place, bacteria are most abundant where plant plankton are also most abundant.
- In addition to bacteria, yeasts are found in the sea, along with molds, bacteriophages, and even some filtrable viruses. Of course, the usual planktonic plants and animals also occur, but ZoBell believes that "the major domain of the marine microbiologist is the community of microbes that, owing to their size and behavior, are studied primarily by cultural methods, or with a microscope at a magnification of about 1,000 times."
- Most marine bacteria are beneficial, but there are a few

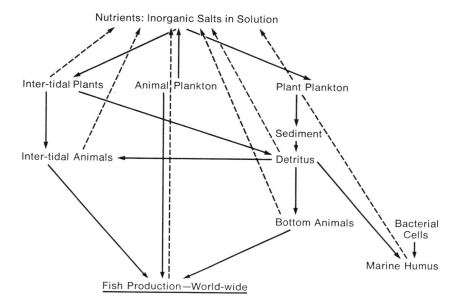

A simplified diagram of the principal relationships among plants, animals, and bacteria in the sea. The dotted arrows show the direction of decomposition of organic matter into inorganic nutrient salts.

that are harmful. Some cause diseases of lobsters, oysters, salmon, and other fish; one causes the wasting disease of eelgrass; and still others cause rapid spoilage of fish and other seafoods after they have been caught.

• Most of the bacteria that cause human diseases (the pathogens) die rapidly in the high salinity of sea water.

• By far the most important activity of bacteria in the sea is the breakdown of organic matter into inorganic plant nutrients. The remains of plants and animals are decomposed as the bacteria assimilate carbohydrates, lipids, proteins, and hydrocarbons, and liberate their chemical constituents as simple, soluble inorganic ions and gases that then become the essential food for the plants of the plankton.

Although the number of scientists who are primarily interested in marine bacteria is now growing at a greatly accelerated rate, there is still no lack of rewarding areas in which they are hard at work, and the more they learn about problems such as the following, the more confidence we can have in man's relation to the oceans of the future.

— How marine microbes fix nitrogen from the air.
— How they infect marine plants and animals.
— How marine algae are affected by viruses.
— How marine microbes produce substances that promote or hinder growth.
— How they modify marine food products.
— How they affect the nutrition of marine animals.
— How they break down natural debris.
— How they concentrate trace elements such as vanadium, manganese, and radioactive isotopes.
— How they affect the characteristics of different kinds of water masses.
— How they contribute to the total mass of living materials.
— How susceptible they are to variations in salinity, temperature, and pressure, and how these influence their race histories.

We see that these are very large problems indeed, especially when we recall that answers to them must be sought over 72 per cent of the earth's surface (about 143,000,000 square miles of area and 327,000,000 cubic miles of volume), at depths nearly as great as seven miles, and at hydrostatic pressures as great as nine tons per square inch.

The best hope of finding useful answers to problems of such great scope lies in international cooperation, and it is indeed fortunate that we can look forward to this with some confidence.

When I went to ZoBell's office on a November morning in 1972, he had just completed a long-distance call that had lasted more than an hour. He sat down in one of the chairs that ring his large conference table and relaxed for a moment, for his telephone is mounted high on the end of a tier of bookshelves, and he stands to use the instrument.

Smiling, he said, "A long phone call. That was Washington. It seems that it is now certain that an international study on pollution of the ocean will get under way. I have just been asked to serve as the marine microbiologist on that study."

He was obviously much pleased, for he said later that although he had been studying marine microbes for more than forty years, he still felt that man's knowledge of the subject was only at the threshold.

As a member of an international study group, he will be ideally placed to advance that knowledge on a world-wide front.

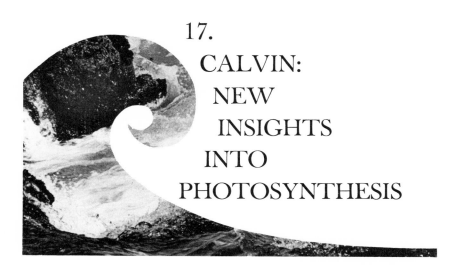

17.
CALVIN: NEW INSIGHTS INTO PHOTOSYNTHESIS

You will recall that the photosynthetic reaction—by which green plants convert the radiant energy of sunlight into chemical energy—has been called the key chemical reaction of the universe. You will also recall that by the middle of the nineteenth century it became possible for scientists to summarize the fundamental transformation in the form of the simple equation:

$$CO_2 + H_2O \xrightarrow[\text{green plant}]{\text{light energy}} (CH_2O) + O_2$$

which is simply to say that through the agency of green plants sunlight transforms carbon dioxide and water into carbohydrate and free oxygen.

The very neatness of this concept suggests that it may be seriously deficient as an explanation of this great energy-trapping process, and that has turned out to be the case. Developments that have taken place since the middle of our own century have opened up vast new areas of understanding about it. Some of these areas may seem to be of purely theoretical interest—at least as far as we can see now; others are of the greatest practical potential for the future good of mankind. At present the whole field is in such a rapid state of flux that it would take a rash prophet to say what development will be of the greatest significance ten years from now. With such reservations in mind, let us look at some of the new knowledge that has been gained about photosynthesis in only the last few years. These are developments that would have been impossible at any earlier time, for earlier science lacked the tools that finally gave the answers.

Let us look first at the part of the green plant in which photosynthesis is carried out. To begin with, it takes place inside the plant cells, and this has long been known. It was also well known that chlorophyll, the green coloring matter of plants, is not uniformly distributed throughout the plant cells, but rather is localized in small bodies within the cells called chloroplasts. Each chloroplast, moreover, is an extremely complex body, as can be seen under the electron microscope. Inside the chloroplasts are thin membranes that stretch from wall to wall, these membranes thicken and darken in areas called grana, and each of these grana may contain from 250 to 300 molecules of chlorophyll. It is at these specific sites that photosynthesis takes place.

Thus one of the tools that has been tremendously helpful in understanding the structure and function of plant cells has been the modern microscope. Of course, we know in a general way how powerful modern microscopy is, but such great strides have been made in recent years that it seems well to be quite explicit about them.

It has been barely three hundred years since there was such a thing as a microscope. When Anton van Leeuwenhoek focused one of his little glass lenses on a drop of stagnant water in 1675 in Delft, he became the first man to see a living creature smaller than those visible to the unaided eye. When Leeuwenhoek died in 1723, he left 247 finished microscopes and 172 lenses, a number of which he willed to the Royal Society in London. These instruments had magnifying powers of from 50 to 200 times.

The scientists who followed tried to increase the magnifying power of microscopes even further, but for a long time they were frustrated by the fact that as they increased the size of the image, they also intensified the color rings that blurred the focus. Then in 1830 the English optician Joseph Lister designed a microscope that eliminated the color rings, and by 1878 the German physicist Ernst Abbe began a series of improvements on this "achromatic" microscope that ultimately led to optical microscopes capable of revealing clear images at magnifications of 2,000 times. That was the limit for microscopes that use ordinary visible light.

To obtain greater magnifications it was necessary to turn to forms of magnetic radiation having wavelengths shorter than light, and streams of electrons were the logical choice for this. Instead of glass lenses, magnetic condensers, objectives, and projectors were used to project the images onto either a fluorescent screen or a photographic plate. James Hillier and Albert F. Prebus of the University of Toronto built an electron microscope that gave clear images at magnifications of 7,000 times. That was in 1937, and before another fifteen years had passed, magnifications as great as 2,000,000 times were possible.

But even that was not the limit. Electron streams do not have the shortest known wavelengths; wavelengths associated with ions are even shorter, and in 1955 Erwin W. Mueller of Pennsylvania State University stripped charged ions from a very fine needle point, and his "field ion microscope" produced images magnified 5,000,000 times.

Thus in the three centuries since 1675, great strides have been taken in shorter and shorter periods of time. We can make the point more vivid by putting the numbers into a little table.

APPROXIMATE DATE	INTERVAL, YEARS	POSSIBLE MAGNIFICATION	INCREASE IN MAGNIFYING POWER
1725	–	200	–
1900	175	2,000	1,800
1937	37	7,000	5,000
1950	13	2,000,000	1,993,000
1955	5	5,000,000	3,000,000

This is a typical example of an exponential rate of progress, and I have emphasized it to show the tremendous strides that microscopy has made, and also because many other tools that today's scientist uses in his routine daily work are comparably more powerful than anything that went before. Not to belabor the point, scientists now have at hand tools so powerful as to have been quite beyond imagination only a few short years ago.

When it became clear that the rather simple chemical compound chlorophyll was essential to photosynthesis, scientists were at once led to try to bring about artificial photosynthesis in the laboratory, using solutions of the green pigment itself, extracted from plant leaves in various ways. None of these worked, however, and it soon became apparent that a great many other conditions above and beyond the mere presence of chlorophyll, carbon dioxide, and water were necessary for the photosynthetic reaction to proceed.

If chlorophyll itself was not sufficient, then the next obvious thing to try was the chloroplasts that contain the chlorophyll. These, as we have already seen, are tiny bodies within individual plant cells in the shape of disks that may contain several hundred individual chlorophyll molecules. The electron microscope shows them to be made up of tightly stacked membranes so arranged that layers of chlorophyll molecules are sandwiched between layers of protein.

In view of such complexity of structure (and no doubt of function), it is hardly surprising that chlorophyll alone would not bring about the photosynthetic reaction. However, a close approach was achieved by Daniel I. Arnon, a cell physiologist working at the University of California in Berkeley. In 1954 he took ordinary spinach leaves, and by the most delicate of disruption and separation processes, was able to liberate undamaged and fully active chloroplasts from the spinach cells. In the presence of these chloroplasts, Arnon was able to carry out the reaction of photosynthesis in the laboratory just as well as the plant itself can in the open field.

Such chloroplasts—isolated from their parent cells—could be regarded as so many tiny chemical factories, busily transforming carbon dioxide and water into carbohydrates and free oxygen. Again, however, this was simply the over-all reaction. Scientists could do little more than speculate about all the individual processes that went on inside these marvelous little factories.

This was even more complicated, and before further insight could be gained about the individual chemical reactions that occur during photosynthesis, more new tools were necessary, and especially the imagination to apply such tools to the elucidation of the photosynthetic reaction. The tools became available shortly after the end of World War II, and the imagination was supplied by Professor Melvin Calvin and his colleagues in the Department of

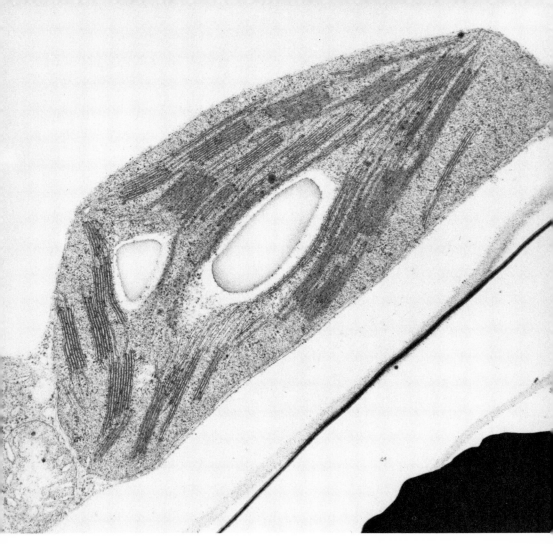

An electron micrograph of a chloroplast from a bean leaf. The small bodies that look like threads lined up side by side are the grana, which contain alternating layers of protein and lipid.

Chemistry and the Lawrence Radiation Laboratory of the University of California at Berkeley.

The key tools were the nuclear reactor, which could produce radioactive isotopes of various chemical elements in practical quantities at reasonable cost, and paper chromatography, a powerful method for analyzing tiny amounts of highly complex mixtures of organic compounds.

The pioneer among the nuclear reactors was the famous Chicago pile of Enrico Fermi, in which, on December 2, 1942, the first self-sustaining fission reaction was made to proceed. Within less than twenty years, newer and far more sophisticated reactors had been built, and they were used to bombard ordinary compounds of phosphorus, sulfur, hydrogen, potassium, sodium, iodine, iron, and copper with neutrons, and to produce a great variety of radioactive compounds that were of unprecedented value both in science and in industry. These isotopes opened up a whole new field in tracing the exact courses of a variety of chemical, biological, and technological reactions.

Plant physiologists, for example, had long yearned to know just what course phosphate fertilizers took in the growth of cereal grains from the sprouting seedlings to the harvested grain. Now, whether in the field or in greenhouses, they could follow the activity of compounds tagged with radioactive phosphorus. One had only to add a bit of phosphate tagged with radiophosphorus-32 to a nutrient solution, and then measure—in later samples of leaf and stem and grain—amounts of phosphorus so small as to be completely beyond detection by any previously known means.

Of all the radioisotopes that the reactors produced, radiocarbon-14 was of the greatest interest to biochemists studying life processes, for the element carbon is the fundamental building block of all organic matter. What made carbon-14 so unusually useful was the fact that it has a half-life of some 5,000 years; thus experiments with it can be carried out without haste, and with as great precision as an investigator may wish. Thus by 1950 Melvin Calvin and his co-workers at Berkeley could obtain—at reasonable cost—as much carbon dioxide labeled with carbon-14 as they would need to carry out a complete study of the path that carbon takes in photosynthesis. Professor Calvin tells me that the first supplies came from the reactor at Oak Ridge, and that larger amounts later came from Hanford.

The other tool available to Calvin can almost be said to be the more spectacular. An analytical method called paper chromatography, it was first developed by the British biochemists A. J. P. Martin and R. L. M. Synge for the separation of complex mixtures of amino acids (the building blocks of proteins) into their several components. This method required only the tiniest samples for the analysis of even the most complex mixtures.

The principle on which paper chromatography depends is that when a few drops of a mixture (say, of amino acids) are deposited onto a large sheet of filter paper, and the paper dampened with various solvents, each amino acid will travel at a

The radioactive carbon-14 that Melvin Calvin used in tracing the path of carbon in photosynthesis was produced in this graphite reactor at Oak Ridge, Tennessee. The world's first full-scale reactor, it began operation late in 1943, and served scientists for two decades as a major research tool. It is now a national monument open to public view.

characteristic rate down the paper. Let us say that a rectangular sheet of filter paper is dosed with an unknown mixture, and then hung with a short side down from a shallow trough containing, say, solvent X. As the solvent spreads across the surface of the filter paper by capillary action, it carries the amino acids down the length of the paper at varying characteristic rates. If the paper is then removed from contact with the solvent and dried, a rough separation of the amino acids will have taken place. This will be evident as a series of discrete spots distributed along the length of the paper. But that is only the first separation. Each of these spots may contain several different amino acids (though the number will be significantly smaller than that of the original sample).

Note, however, that the resulting dried filter paper may now be turned at right angles and suspended from a trough containing a *different* solvent Y. The spots from the primary separation will now again travel at different rates under the influence of the second solvent, and when the sheet from that operation is dried, each individual amino acid will occupy a characteristic position upon the two-way map of the filter paper.

Perhaps a crude analogy will help to make this clear. Imagine an ordinary checkerboard in place of the filter paper, and for the purpose of this illustration, place it flat on a table. Then imagine that instead of a few drops of an unknown amino acid solution we have a teaspoonful of tiny beads of many different colors. These we will place in a little pile on the lower-left-hand square of the checkerboard. Let us suppose further (and only for the purpose of this analogy) that if we direct a gentle stream of air up the left-hand row of the board, we can separate the original teaspoon of beads into eight different groups occupying each of the eight squares along the left-hand row.

Now let us turn the checkerboard at right angles, so that what was originally the left-hand row is now at the bottom of the board. Again we direct our imaginary stream of air across each bottom square; this will arrange the beads from each of these squares from the bottom to the top of the new position of the board. Since there are sixty-four squares on a checkerboard, we should be able to separate our original pile of beads into up to sixty-four groups having sixty-four different characteristics.

So much for the analogy. In actual practice, in which large sheets of filter paper were used with two carefully chosen solvents, as many as 526 amino acids of 18 different types have been accurately separated—that is, 526 different spots each in its char-

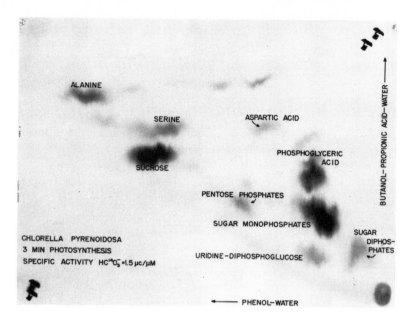

An autoradiograph showing the individual chemical compounds produced by Chlorella *during a three-minute exposure to carbon dioxide tagged with carbon-14.*

acteristic location on a filter-paper map measuring about eighteen by twenty-four inches!

But even such a visual identification of spots does not exhaust the extreme delicacy of the method. In addition to the ability to separate and identify by sight such a bewildering array of different compounds, it is also possible to detect their presence when they occur in such trace amounts as to leave no significant spots at all. Compounds tagged with carbon-14 are radioactive and will leave a mark on a sheet of X-ray film. It is only necessary, then, to lay a paper chromatogram over X-ray film, allow exposure to proceed for some hours, and develop the film in the ordinary way. The resulting images on the film will be many times more sensitive than the chromatogram itself. Such a picture is called an autoradiograph.

These tools, then, coupled with a highly imaginative experimental approach, provided Calvin and his colleagues with the means for probing the innermost secrets of the single cells of green plants.

Calvin was already well aware, as we have seen, that each such single cell may contain a few hundred chloroplasts, that these are the seats of the photochemical reaction, and that their structures and functions are exceedingly complex. He had ready at hand an adequate supply of carbon dioxide labeled with radioactive carbon-14 dioxide (let us now write it $C^{14}O_2$); and he knew that he could separate and identify a very large number of different organic compounds through which the C^{14} might pass before it reached a storage point in the plant in some form of carbohydrate.

In his own reports, which extended over a number of years and to which many co-workers contributed, Calvin says that the design of the experiments was quite simple. He took small samples of the one-celled green plant *Chlorella* (grown under the most carefully controlled conditions to ensure that each sample would be like every other sample), fed them with $C^{14}O_2$ for varying periods of time, and determined which compounds appeared at the different times of exposure. As he shortened the exposures to periods of fractions of a second, only those reactions that took place the earliest could then have occurred, and these turned out to be the formation of high-energy phosphate compounds—compounds capable of driving the reactions vigorously during the early stages of the cycle.

As Calvin lengthened the times during which his *Chlorella* cells could take up $C^{14}O_2$, the number of different compounds increased until the entire sequence had run full cycle, and newly

A suspension of the alga Chlorella *was briefly exposed in bright light to carbon dioxide tagged with radiocarbon-14. A sample of the exposed algae was then killed by dropping it into alcohol in the flask seen at the right of the clock.*

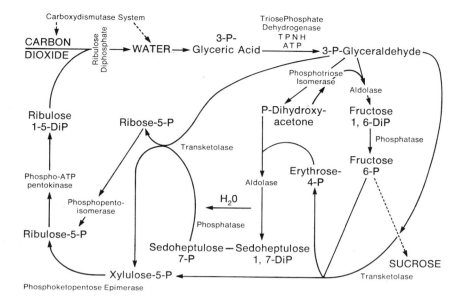

This diagrammatic sketch shows the path that carbon takes in the photosynthetic transformation of carbon dioxide to the sugar sucrose. Note that this cycle is only one phase of the photosynthetic process. Not shown is the release of free oxygen or the synthesis of fats or proteins.

admitted $C^{14}O_2$ began to react with the end products of the cycle. Thus at each turn of a cycle a new increment of $C^{14}O_2$ was taken up.

The over-all pattern—or sequence of reactions—is now widely known as the Calvin cycle, and most biochemists agree that it is the most likely explanation that we have of the path of carbon in photosynthesis. For this contribution Calvin was awarded the Nobel Prize in chemistry for 1961.

The accompanying diagram is an extremely sketchy representation of the reactions that a carbon atom participates in on its way from the carbon dioxide (which you or I may have exhaled) to the sugar (which may have come from beets or cane) which we now find on our table. Unless you happen to be a specialist you will not want to trace out these reactions in detail, and I present the diagram for two other reasons. The minor one is to show how inadequate the equation

$$CO_2 + H_2O \xrightarrow[\text{green plant}]{\text{light energy}} (CH_2O) + O_2$$

that we started out with really is. The major one is to demonstrate how the laboratory methods that have become available in just the last few years have outstripped everything that went before.

And that is not all by any means, for it is only the beginning of our understanding of photosynthesis. In addition to the carbohydrates (such as the sugar of our example), fats are also produced by photosynthesis, as are also the amino acids from which proteins are built up.

Any detailed consideration of these reactions would carry us much deeper into biochemistry than we have any need to go. It will be enough for our purpose to realize that science now knows enough about the reactions by which plants produce carbohydrates, fats, and proteins, so that *control of these reactions now seems within reach.* In a paper that they presented at an international symposium on photosynthesis in plant life in October 1966 in Chicago, James A. Bassham and Richard G. Jensen, two colleagues of Melvin Calvin's, pointed out that a better understanding of the mechanisms that regulate the paths that carbon takes during photosynthesis should be of value in future agricultural research. "Perhaps it will be possible," they said, "to manipulate the regulatory mechanisms to provide more or better end products in photosynthesis in green leaves. . . .

"The importance of such control of products of green leaves must not be underestimated. Leaves which can be consumed directly as food by people may well become much more important in agriculture in the future. If leaves used for fodder are enriched in fats and proteins, greater productivity of animal protein can be achieved. Utilization of such potentially vast crops as jungle foliage might become an economic reality if a light chemical spraying by airplane could result, a few days or weeks later, in leaves greatly enriched in fats and protein.

"Such hopeful predictions are mere speculations today. Nonetheless, today's and tomorrow's discoveries of the nature and distribution of photosynthetically reduced carbon, and the mechanisms of regulation of this distribution, provide a basis for a new era of agricultural experimentation."

What seems worth emphasizing above all else is that only within the last few years has it been possible to make statements of such new optimism concerning the future of man's nutritional well-being. To that we may add that each new discovery revealed to us strengthens our reasons for believing that deeper and deeper understanding of nature's innermost secrets lie close at hand.

"But," cries a skeptic, "we have no evidence that man will apply technological advances to the welfare of mankind. On the contrary, we have abundant evidence that the present preoccupation with technology is precisely what is dehumanizing the human condition. Scientists have no concern for the effects of their work upon the welfare of humanity."

This has come to be a kind of cliché among those who classify any kind of scientific or technological progress as inherently bad. To give them their due, we must admit that over the course of human history many scientific and technological advances *have* been misused or perverted. But to maintain that this reflects anything about the motives of scientists or the objectives of science is simply to distort history.

What, then, *do* scientists believe? What *does* motivate them? Why do they spend endless hours over the most tedious details of laboratory routine—and more endless hours at the study lamp, painfully deciphering the reports of other scientists writing in other languages? Answers of many different sorts have been advanced, but to me one of the most recent statements is one of the most revealing. Melvin Calvin made it when he recently summed up his work on the origin of living systems on the earth and elsewhere—a study closely related in chemical principle to the photosynthetic reaction that we have just considered. Under the heading "Personal Experience" in *Chemical Evolution* (1969), Calvin wrote:[1]

> I now come to the last element in this structure. Following this attempt to draw from the scientific experience some generalized concepts, it seems worth while to relate some personal history.
>
> The fundamental conviction that the universe is ordered is the first and strongest tenet. As I try to discern the origin of that conviction, I seem to find it in a basic notion discovered 2000 or 3000 years ago, and enunciated first in the Western world by the ancient Hebrews: namely that the universe is governed by a single God, and is not the product of the whims of many gods, each governing his own province according to his own laws. This monotheistic view seems to be the historical foundation for modern science.

---

[1] From *Chemical Evolution: Molecular Evolution Towards the Origin of Living Systems on Earth and Elsewhere*, by Melvin Calvin. Copyright © 1969 by Oxford University Press. Reprinted by permission of the publisher and the author.

A second tenet seems to lie in a fundamental quality of the human mind, which has been built into it by natural selection over the millennia of its evolution, namely, the need to know and understand. How this need to know and understand was evolved is probably traceable to some survival-selection pressure in the primitive being that gave rise to mankind. Be that as it may, it is here today, and has been the source of man's greatest achievements in all the areas of his activity: religion, the arts, the sciences, and the like.

When this "need to know" is coupled with the conviction that it is worth knowing—that the search, in principle at least, is leading towards an understanding of an unattainable infinity of knowledge that is the law of the universe—a code of conduct for the hour, the day, the week, the year, the life, comes forth.

When one is discouraged with the mass of scientific or sociological data and the need to encompass and comprehend all that man knows already, it is time to sit back and meditate on the simple one-celled amoeba. The amoeba has only one cell, but it can make itself x times over in an hour; it can cause untold trouble for man, or many other creatures of nature, by its capacity to interfere with myriad cells. It has only one cell, yet it changes its shape, its size, its location, and, without willing it, it changes its progeny. If an amoeba, with a single cell, can do so much, there is room yet for wisdom in man. Man will always pursue; he will always find; and he will always be different tomorrow from what he is today.

Man has been learning slowly for centuries that his own life depends to a great degree on the life of his neighbor. For centuries this process was snail-paced, as when his own sewage was thrown into the street for his neighbor to walk on. Eventually, in self-defense, a community system was devised. Today, every car's exhaust contributes to the irritation of cancer-susceptible cells in each of us.

Once man struggled for the education of only his own offspring. But now one must be aware that the world his child will inherit is only so good, or so safe, as the least-educated child. For the educated man can be destroyed, or mutilated, or enslaved by the man who has no education—no education from home, from school, from community. This man, who has nothing to lose, does not fear to destroy. For 2000 years religious precepts have taught that man must be his brother's keeper, but it remains for science to give example after example of the truth of this early philosophical concept.

There can be no ultimate right, no final understanding, no permanent solutions for the problems of mankind. For change is inherent in the structure of the molecules of which we are composed. This is perhaps the hardest truth, for it allows no rest.

The slogan of yesterday, "for God and Country," is still the star of today, only the meanings of the words have changed: "God" is "the Universe" and "Country" is "the Human Race."

18. ON SEEKING A POINT OF VIEW

Some deeply concerned biologists have called the late 1960's and early 1970's "the age of eco-hysteria." They believe that much of what is being written and published about the condition of our environment is deliberately overstated, and that by diverting our attention from the very real problems that do exist, it is doing much more harm than good. A glance at any newspaper or magazine file for the period will show the cause for their concern.

An Associated Press story out of Washington released for publication on October 18, 1971, carried the headline OCEANS NEAR DEATH, SAYS COUSTEAU. "Jacques Cousteau, the world's most famous undersea explorer, predicted today the death of the world's oceans within the next half century unless industrialized countries 'tame competition' and cooperate to save them. . . . 'People have thought the legendary immensity of the ocean was such that man could do nothing against such gigantic force,' Cousteau said. 'Well, now we know . . . what we are facing is the destruction of the ocean by pollution and by other causes. . . . If nothing was done today I would say that maybe 30, 40, or 50 years could be the end of everything.' "

Life for September 24, 1971, carried a story by Galen Moore under the headline THOR HEYERDAHL'S PAPER BOAT, PLOWING A FILTHY OCEAN. "On 43 out of 57 days at sea, [Heyerdahl found that] the ocean was visibly polluted. 'If that much pollution was visible,' Heyerdahl exclaimed, 'just think how much more there must have been that we didn't see. I started out this voyage to get a glimpse of man's past, but I got just as much of a glimpse into man's future. I was scared.' He decided to dedicate every spare moment to alerting the world."

A United Press International story out of New York appeared under an October 11, 1971, headline, ECOLOGISTS MAKE USE OF OVERSTATEMENTS. "An earnest political ecologist told a women's club luncheon meeting, 'Lake Erie is dead, no fish, no microscopic life, no nothing.'

"The club ladies were impressed, even shocked, but Ted Fearnow, company biologist and land manager for International Telephone and Telegraph's Pennsylvania Glass Sand Co. at Berkeley Springs, W. Va., reading about it in a local paper, was suspicious. He telephoned the New York State Fisheries Bureau.

" 'Lake Erie is filthy, all right,' he was told, 'but it still yields sixty million pounds of commercial fish a year.'

"This is part of a phenomenon which caused the *New Yorker* magazine, which sparked popular excitement over ecology by pre-publishing Rachel Carson's *Silent Spring*, to remark recently that 'in the war strategy of the conservation movement, exaggeration is a standard weapon.' "

What is the truth? Is calamity indeed upon us? Or are we being taken in by deliberate and calculated exaggeration? And if we are, what harm is there in it? Perhaps it takes calculated overstatement to awaken a lethargic public consciousness.

The harm, it seems to me, is very real and very frightening. Aside from intellectual dishonesty (which must certainly say something about our moral values), overstatement plays such havoc with our emotions that we tend to react in illogical and unconstructive ways.

John Maddox, editor of the prestigious British journal *Nature*, has put the case cogently in his recent book, *The Doomsday Syndrome*. Reporting on the book, *Saturday Review of the Society* said, "The environmentalists, a leading British scientist charges, may be the most insidious of all plunderers of our planet. Using 'a technique of calculated overdramatization,' they have deflected attention from the genuine ecological issues we face and blinded us to solutions that exist now."

Thus we face a problem (or rather a whole array of problems) of extraordinary difficulty and complexity. As responsible citizens, we must take some point of view—if not to urge it on our elected representatives, then at least to express it to friends and acquaintances who value our opinions.

And after all, the way toward finding such a point of view may not be too difficult, if we break the problem down into small component parts and carefully seek answers to these parts. Let us consider a specific example.

From several of the preceding chapters in this book, we already possess a substantial background of information about photosynthesis. We know that it is the chemical-biological reaction that produces all the food that is consumed by all the creatures on earth. Perhaps we can use that background to explore a critical question that certain environmentalists have posed. Let us put it in the form of the unequivocal statement that they have made. *The world has run out of food. Unless the growth of world population is immediately checked, mankind will starve within the next decade or two.*

Is this a valid warning that we should seriously heed, or is it an example of calculated overdramatization?

The emotional overtones of statements like this present such serious threats to rational judgment that we will do well to look briefly at the kind of thinking that such problems call for. In essence, this thinking is simply the logic of the ancient Greeks, but it seems to take on new life and meaning when it is restated in terms related to today's problems. Monroe C. Beardsley has done this brilliantly in his little book *Thinking Straight,* which I would heartily recommend. In his "Preview," Beardsley introduces the techniques of thinking straight about current problems.

> One of the difficult issues confronting our society today concerns the use of certain chemical compounds known as pesticides in controlling undesirable animals and plants. Since the introduction of DDT in 1942, the use of these chemicals has increased enormously; today there are more than 50,000 varieties, of which over a billion pounds are produced each year. Those who believe pesticides to be one of the greatest inventions of man point out that without them, production of

apples, citrus fruits, cotton, and potatoes would be 50 per cent less—that before their wide use, blight destroyed half the crop of tomatoes in 1946, and in other years the corn borer ruined millions of acres of sweet corn—that DDT has virtually wiped out malaria and typhus in this country. On the other hand [some] conservationists and biological ecologists point out that millions of birds and fish have been killed off by heavy use of pesticides—that these compounds build up in the fatty tissues of the body, where their ultimate effect is unknown—that allergic reactions to them may be increasing as our diet contains more and more of them, since once they are sprayed on a plant, they are nearly impossible to dislodge.

As public debate widens and grows more intense, the ordinary citizen finds it difficult to get his bearings on the matter. He does not know how to vote, or what to write his senator, or whether to support or oppose his local borough council when it is considering a campaign of spraying mosquitoes. In the emotional atmosphere, there is danger that an excessive fear of consequences will lead to unwarranted restrictions on the use of pesticides and on the development of further types. On the other hand, there is also a danger that ignorance or apathy, or the propaganda campaigns of the pesticide industries, will make us think that all is well, and that ten or twenty years from now, when the long-range consequences become clear, our children will curse us for spreading these chemicals through our topsoil and our rivers.

It is a situation that calls not for hysteria or for smug complacency, but for *thinking*.

In a popular sense, almost anything that passes through the mind can be called thinking—including memories and daydreams. But I am using the word in a narrower sense. Thinking is a series of ideas that is directed, however vaguely, toward the solution of a problem. When what goes on in our minds is initiated by some question that we want to answer, then it takes on purpose and direction: it has something to aim at, and its course is under the direction of the question that started it off. Pesticides are a problem, undoubtedly. This is the *practical* problem: What are we going to do about them? But this also involves a *theoretical* problem: Will the controlled use of pesticides produce benefits without corresponding evils? As soon as we ask ourselves this question, all sorts of leads suggest themselves, for there is much that we will want to know: about the chemistry of the compounds, about their immediate and long-range physiological effects, about the balance of nature, and so on. But even if the inquiry takes us far afield, we will have a standard of relevance, for we will be guided all along by a desire to find what we need in order to answer the original question.

What thinking is after is simply the truth. When it results in truth, the thinker acquires new knowledge. That is the measure of its success. That is what makes thinking *good* thinking.

When we consider the nature of thinking more closely, as it might develop, for example, in the pesticide problem, we find that it has two fundamentally different aspects: a *creative* aspect and a *critical* aspect.

Creative thinking gives us new ideas. At some point—and it is to be hoped at many points—along the line, someone will come up with ingenious suggestions for new experimental studies. For example, what is the best way to find out how much pesticide material there actually is in the food we eat? And, again, how can we predict the long-range effect of pesticides, without waiting around until it is too late? Moreover, some of those working on the problem will come up with original ideas about possible alternatives to the use of pesticides. These may be radical, yet they may be worth trying; for example, importing or breeding special insects to attack the ones we want to get rid of, or killing the pests with high-pitched sounds or luring them to their death with artificial female-insect scents.

This kind of thinking is an exercise of the imagination, which is needed in biology and engineering just as it is in writing poetry or music. So one task before us, if we are interested in good thinking, is to discover how we can make thinking more creative. Unfortunately, we do not yet know very much about this matter. Psychologists have studied creative thinking, and we can say something about the personality factors and environmental factors that seem to favor or hinder it. And teachers of literature have long contended that one of the values of the arts, especially of literature, is to enlarge and free our imaginations. But there is no handy set of rules to follow to make our thinking more creative. Perhaps there is even something a little paradoxical in the very notion of rules for creative thinking—since this is thinking that leaps out in new directions to unexplored territory, and moves in sudden, unexpected, and unpredictable ways.

Critical thinking comes into play after we have an idea to try out, a theory to test, a proposition that someone wants to prove or to refute. Are we sure that all the evils sometimes attributed to DDT really are due to it? And what about the benefits? Granted that between 1940 and 1962 the average farmer increased his productivity from providing enough for himself and ten others to providing enough for himself and twenty-six others—how much of this increase is due to pesticides and how much to other factors? Whenever a claim is made to knowledge, there is an occasion for critical thinking —that is, for a careful and serious effort to test that claim, as far as possible.[1]

Some of the environmentalists, then, have presented us with plenty of occasions for critical thinking—for careful and serious efforts to test their claims. Let us consider just one of these, the claim that unless the growth of world population is immediately checked, we face starvation.

As our first step, we need to examine very carefully the exact wording in which this claim is made. Probably Dr. Paul R. Ehrlich, professor of biology at Stanford University, is the best-known exponent of this thesis. At least, his book *The Population Bomb* has been very widely read and widely quoted. As of February 1971, it had been through twenty-two printings in its first edition, and was still selling in the revised edition. We will assume that in the "Prologue" of the book Dr. Ehrlich has stated the essence of his case as he sees it. That "Prologue" reads as follows.

> The battle to feed all of humanity is over. In the 1970s and 1980s hundreds of millions of people will starve to death in spite of any crash program embarked upon now. At this late date nothing can prevent a substantial increase in the world death rate, although many lives could be saved through dramatic programs to "stretch" the carrying capacity of the earth by increasing food production and providing for more equitable distribution of whatever food is available. But these programs will only provide a stay of execution unless they are accompanied by determined and successful efforts at popula-

---

[1] Monroe C. Beardsley, *Thinking Straight*, 3rd ed., copyright © 1966. Reprinted by permission of Prentice-Hall, Inc., Englewood Cliffs, New Jersey.

tion control. Population control is the conscious regulation of the numbers of human beings to meet the needs not just of individual families, but of society as a whole.

Nothing could be more misleading to our children than our present affluent society. They will inherit a totally different world, a world in which the standards, politics, and economics of the past decade are dead. As the most influential nation in the world today, and its largest consumer, the United States cannot stand isolated. We are today involved in the events leading to famine and ecocatastrophe; tomorrow we may be destroyed by them.

Our position requires that we take immediate action at home and promote effective action worldwide. We must have population control at home, hopefully through changes in our value system, but by compulsion if voluntary methods fail. Americans must also change their way of living so as to minimize their impact on the world's resources and environment. Programs which combine ecologically sound agricultural development and population control must be established and supported in underdeveloped countries. While this is being done we must take action to reverse the deterioration of our environment before our planet is permanently ruined. It cannot be overemphasized, however, that no changes in behavior or technology can save us unless we can achieve control over the size of the human population. The birth rate must be brought into balance with the death rate or mankind will breed itself into oblivion. We cannot afford merely to treat the symptoms of the cancer of population growth; the cancer itself must be cut out.[2]

A logical first step in analyzing such a statement would be to consider its general tone. What of the choice of words and phrases here? How about the terms "battle," "starving to death," "stay of execution," "affluent society," "standards . . . are dead," "famine," "ecocatastrophe," "destroyed," "value system," "permanently ruined," "oblivion," "cancer"? Emotionally supercharged words indeed. Do they tell us something significant about the author's intentions?

In *Thinking Straight,* Monroe Beardsley speaks of the

---

[2] Copyright © 1968, 1971 by Paul R. Ehrlich. Reprinted by permission of Ballantine Books, Inc. All rights reserved.

"manipulation of emotions as a means of effecting oversimplification or distraction. This device (*the appeal to emotion*) involves the use of emotive language, together (generally) with some reliance or suggestion to hint at connections that we might be skeptical about if they were brought out into the open. Many different emotions can be appealed to. Very often it is fear. Of course there is nothing inherently wrong about spreading an alarm—when there is something to be alarmed about, and when the amount of alarm is proportional to the danger, and when the manner of giving the alarm is such as to evoke rational action rather than blind panic and confused terror."

How does Dr. Ehrlich's statement stand up in that light? You will have to make your own judgment, but mine is that it is more likely to evoke blind panic than rational action.

Such matters of tone, however, are really secondary, although they are surely evocative enough to lead us to want to know more about the facts. Ehrlich has surely put forward a proposition that he very much wants us to accept. Here, then, is an occasion for a careful and serious effort to test his claim.

To do that, we need to cut through the emotive language and restate exactly what the claim is—in language as direct and factual as we can manage.

How about this? *The world cannot produce enough food to support a human population that is increasing at its present rate.* It could be worded in various ways, but that seems a fair statement of Ehrlich's claim. And we should be able to test it, one way or another.

Can we come up with an ingenious suggestion, perhaps, for a new experimental study? Or better yet, has someone already thought up and carried out such a study? Fortunately, as it turns out, someone has. Our problem is to find it.

This calls for a search of the literature—at first glance an appalling task because of sheer volume. But we have some guidelines that will enable us to narrow it sharply, and soon to pinpoint it. Our basic guideline depends on the fact—already well known to us—that all the world's food is produced by photosynthesis, and the obvious place to begin is in books about photosynthesis. We riffle through them, glancing at the chapter titles. Shortly we have a clue. In a book published as recently as 1967, entitled *Harvesting the Sun: Photosynthesis in Plant Life,* we find a chapter called "Photosynthesis: Its Relation to Overpopulation." Now the excitement of the first quick look is over, and we begin to read carefully. At the end of an hour of reading and reflection we know that we have the answer to our question.

Harvesting the Sun turns out to be a symposium—a collection of reports on many different aspects of photosynthesis presented at a meeting of internationally recognized authorities in Chicago in 1966. The author of the paper on photosynthesis in relation to overpopulation was C. T. de Wit, a member of the staff of the Institute for Biological and Chemical Research on Field Crops and Herbage, in Wageningen, the Netherlands. With that background, it seems likely that Dr. de Wit can speak with special knowledge of his subject. We note, moreover, that his paper is a straightforward scientific report, entirely devoid of any attempt to appeal to the emotions. He simply states his facts, the conclusions that he thinks they lead to, and why he thinks so.

He began with the fundamental fact that photosynthesis is the only source of food on earth. He then set out to answer the question "How many people could live on earth if photosynthesis is the limiting process?"

He broke this question down into a series of sub-questions, including the following:

● How many pounds of carbohydrates will an agricultural crop produce per acre-hour of exposure to sunlight of maximum intensity?
● How much difference is there in productivity between clear days and overcast days?
● What differences are there between crops with large horizontal leaves and crops with small leaves pointing in many directions?
● What is the effect of the scattering of the incoming light?
● How do the height of the sun and the condition of the sky affect directional factors?
● How does the latitude affect the rate of photosynthesis?
● How does the average twenty-four-hour temperature affect the rate of photosynthesis?
● How much food per year does a human being need?

With well-established numerical values of all of these factors—under a wide range of circumstances and conditions—Dr. de Wit subjected many different combinations of these to advanced computer analysis. He was so astonished at the result that he made a number of *downward* revisions to be on the conservative side. But even after such revisions he came to "the staggering conclusion that a *thousand billion* people could live on the earth if photosynthesis is the limiting factor!"

He immediately noted that "this is how many could live *from* the earth; not *on* the earth." Nevertheless "how many could live *from* the earth" is the only question that we set out to answer. Pursuing that question further, de Wit concluded: "At present there are about 3 billion men living on earth. The predictions are that there will be 6–7 billion around the year 2000. At this rate of increase, the number of 100 billion will be reached in 200 years." That, be it noted, is just one-tenth of the thousand billion that photosynthesis could support!

Does this answer the specific question that we asked to start with? It seems to me that it does. The world *can* produce enough food to support a mushrooming population for hundreds of years to come. And that particular line of Dr. Ehrlich's argument appears to collapse.

Please note well, however, that this does *not* mean that Dr. Ehrlich is wrong in arguing that limiting the growth of the human population would be socially desirable. He advances many arguments in support of that thesis also. But before we could agree or disagree with any of these arguments, we would have to examine each of them in turn along the same lines as we have done in our example. One of the most important of these would be: "How many people could live and work comfortably on earth if food were not the limiting factor?" That again would have to be broken down into many sub-questions, and you might find it amusing to formulate some of them. We will not take time to do so here.

You will also have noted that we have said nothing about the potential productivity of the sea in relation to the world's food supply. That question is so important and so complex that we will shortly give a full chapter to it.

19. BENSON: ENERGY FLOW IN OCEAN LIFE

Of all the displays of nature, one of the most significant takes place every spring in the Strait of Georgia. Here, in the blue waters between green-carpeted Vancouver Island and the snowy peaks of British Columbia, a fundamental drama of energy transfer among billions of living forms unfolds.

It begins with the May bloom of the plant plankton, mostly the one-celled plants called diatoms—the microscopic algae that have elaborately fashioned shells of silica and chloroplasts whose green chlorophyll is partly masked by a brownish pigment. "When first we see a sample of plankton rich in diatoms," said Sir Alister Hardy, "under the high power of a microscope, it is like looking at a group of crystal caskets filled with jewels as the strands of sparkling protoplasm and groups of amber chloroplasts catch the light."

We see amazing differences in the forms of the individual cells and in the ways that they group themselves together. Here are a couple of dozen lollipops arranged in a circle, with flat disks pointing inward and handles pointing out—rather like an elaborate necklace. There is a group of hourglass figures bent like a horseshoe. Little pillboxes are threaded on a string, or float as free individuals. Some have hairs, or bristles, or spines, and some, shaped like the letter "S," are bare.

Diatoms are also remarkable in the products of their photosynthesis. Rather than manufacturing sugars and other carbohydrates as most green plants do, they produce and store oils or fatty acids. Sometimes the production is so great that they form an oily slick on the surface of the water that may be miles across.

It is the spring chemistry of nature that leads to this tremendous diatom bloom. As the warming sun strikes the mountain peaks, the snow packs begin to melt and torrents of water from the heights rush down the Fraser River and its tributaries to the sea. And with this rush of water come nutrients washed from the soil: phosphates, and nitrates, and especially silica, from which the diatoms build their shells. The Strait of Georgia abounds with mineral nutrients, and the diatoms multiply—as the days grow longer—doubling their number every day or two, until there are tons of them for every acre of open water. At such times and places the diatom populations are far greater than they are elsewhere in the oceans, but you can collect them at any seaside in a fine gauze net. Their variation throughout the seas can be likened to clouds in the air, continuously changing their sizes, shapes, and locations.

In spring the Strait of Georgia provides grazing of the richest sort for the herbivorous animals of the sea. The most conspicuous of these, and the most plentiful, are the little shrimplike creatures known as copepods—so called because they have feet that are shaped like oars. The copepods are the first animals in this marine food chain: they are the predominant animals that graze on microscopic plants, and they in turn provide the major food supply for the smaller fishes, such as sardines, anchovies, herring, and young salmon; but basking sharks and whalebone whales gulp them down in great quantities also.

The copepods are by far the most numerous animals in the food chain of the sea. It is generally agreed that at least 80 per cent of all the ocean animals are copepods. There are more than 6,000 known species, and undoubtedly many more that have never been seen. Known species range in size from a pinhead to an inch or more in length, but those that are most important in the food economy of the sea belong to the genus *Calanus*. These are about

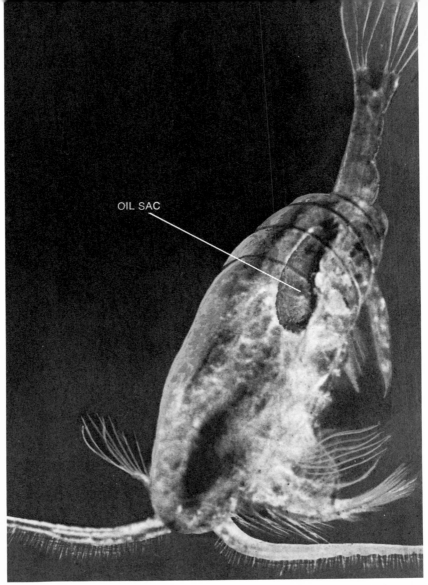

Enlarged photo is of the 1.8-inch marine copepod Calanus, *known as the "insect of the sea." This animal makes wax from oils of algae and stores it in an oil sac at the rear of its body. It uses this reserve as a food supply during periods of starvation and during long winter hibernation.*

the size of a grain of rice. *Calanus atlanticus* is the most important food of the herring, as Sir Alister Hardy showed in his studies of the North Atlantic herring fisheries. Other species, especially *Calanus finmarchicus, Calanus helgolandicus,* and *Euphausia pacifica,* have been the subjects of one of the most recent and most significant studies of the marine food chain and the energy transformations that maintain it.

The principal focus of this advanced ongoing work is at the Scripps Institution of Oceanography in La Jolla, California. There, laboratory work of the most esoteric sort is in full swing, and from there at strategic times sails one of the most sophisticated research vessels afloat—the *Dolphin*.

The *Dolphin*'s chief role is as an aid in the efficient collection of copepods in large numbers for later laboratory study, either of themselves or of their metabolic products; and as an observation station from which to observe the relationships between the diatom bloom, the presence of copepods in great numbers, and the appearance of baby chum and sockeye salmon.

In the spring of 1971, the *Dolphin*—as trim a ninety-five-foot vessel as one might hope to see—was busily at work in the Strait of Georgia. Owned by a San Diego businessman who lends her for research expeditions, she is especially well fitted out for copepod work. This mission, sponsored by the National Science Foundation and the Foundation for Ocean Research, was an

The research vessel Dolphin *on location for the study of the energy relationships among diatoms, copepods, and salmon.*

essential phase of the continuing studies that are going forward at Scripps on the metabolism of fats and waxes in the diatom–copepod–salmon-fry food chain.

In May each year, the copepods end their winter-long hibernation near the bottom of the cold, dark waters of the British Columbia inlets, and rise to within thirty feet of the surface to graze on the spring bloom of diatoms. So May is the ideal time to collect copepods in great numbers, principally to accumulate a supply of the waxes that they contain. These waxes will be used later to study the nutrition of copepods in captivity, as well as that of other marine animals.

At first glance, one might well wonder how large expenditures of time, talent, and money can be justified in studying mere copepods—animals that most people have never heard of, and fewer yet have ever seen. Three basic facts will help to make this clear. First, fats and waxes are essential to animal nutrition, and Scripps scientists have estimated that off the coast of California alone, the copepod population contains some 800,000 tons of liquid waxes. Second, some *half of all the photosynthetic products of the world's oceans is converted at one time or another by copepods into liquid wax,* and we know that photosynthesis is the source of the world's life energy. Third, Scripps scientists believe that new insights into how copepods (as well as fishes) handle fats and waxes in metabolism will lead to new understandings of critical factors in human metabolism—especially the role of cholesterols and of the fatty aldehydes.

Quite obviously, then, we move from what at first appeared to be an idyllic springtime adventure in collecting some curious sea creatures to the very forefront of research in the life sciences—research that may well affect us all.

Before we proceed further, however, a word of warning is in order. Many of the investigations that we have seen so far in this book have been essentially finished pieces of work. Not that they were not subject to later refinement, or modification, or reinterpretation; but they were, as it were, all of a piece. Thus when Priestley and Ingenhousz had done their work, it was known once and for all that plants consume carbon dioxide from the air, release oxygen to it, and that sunlight is necessary for the process. When Robert Mayer had heated a gas in a cylinder with a piston atop it, and measured the amount of heat needed to raise the temperature of the gas by a given amount both when the piston was held stationary and when the gas was permitted to push it up, the mechanical equivalent of heat was known for all time.

Even when Melvin Calvin traced out the complicated path that carbon takes when a plant converts it from carbon dioxide to sugar, that pathway became known, and scientists could use it as a steppingstone to further understanding of other reactions that take place during the photosynthetic process.

The copepod studies are not yet wrapped up quite so neatly. The investigations are going forward along many different lines, and new ideas for new lines are being generated almost daily. Thus the story that we have to tell is a moving, flowing one. That is just what makes it so stimulating and exciting. With no apology at all for the fact that we shall find few final answers here, let us look at this science on the move.

The first thing that we note is the very wide range of scientific disciplines that is being brought to bear on the problem. Here is no nuclear physicist, for example, trying to figure out a magnetic bottle in which a thermonuclear reaction may be contained. Here, rather, are specialists from many different fields, and with wide-ranging backgrounds of experience, all applying their individual skills toward one aspect or another of the problem of how copepods accumulate, transfer, and store energy.

A word about the backgrounds of the men who are attacking the problem will indicate the breadth of their capabilities. The principal investigator and leader of the group is Professor Andrew A. Benson, whose unusually broad training and experience singularly qualify him for this responsibility.

The keys to understanding the basic changes of matter and energy that occur within a marine food chain are to be found in the fundamental sciences, and Dr. Benson holds his degrees in one of the most basic of them—organic chemistry. In 1946 he was named assistant director of the bio-organic research group at the Lawrence Radiation Laboratory in Berkeley. In that position he contributed extensively to the work that Melvin Calvin and his group were doing on the path that carbon takes in photosynthesis, including the identification of nucleotide coenzymes, studies of amino acids, and studies of how radiocarbon is distributed in plant seedlings. These investigations required the analysis of the most minute amounts of complex plant materials, and Benson became intimately familiar with the sensitive and powerful methods used by Calvin's group.

In 1955, Dr. Benson became an associate professor of agricultural and biological chemistry at Pennsylvania State University and advanced to full professorship in 1959. In the years 1961 and 1962 he was professor in residence in physiological chemistry and biophysics at U.C.L.A. and research biochemist in the Laboratory of Nuclear Medicine. In 1962 he was called to Scripps Institution of Oceanography as professor of biology, carrying with him honors that included the Sugar Research Foundation Award, the Fulbright Lectureship at the Agricultural College in Norway, and the Lawrence Memorial Award of the United States Atomic Energy Commission. More recently he has been director of the Physiological Research Laboratory at Scripps; in 1973 he was elected to membership in the National Academy of Sciences. In recent years he has turned his attention to the ecology and physiology of copepods.

In contrast to the marine investigations of earlier decades, experts of many different kinds are now cooperating in the copepod investigations, either aboard the *Dolphin* or in the well-equipped Physiological Research Laboratory at La Jolla.

Among these are Dr. Judd C. Nevenzel, associate research biochemist at Scripps; Dr. Osmund Holm-Hansen, research oceanographer from the Institute of Marine Resources; Donald A. Wilke, director of the Vaughan Aquarium-Museum; Jed Hirota, graduate student in biological oceanography at Scripps; and Bette Baker, Scripps technician. Other institutions are represented by Dr. Nestor Bottino, professor of biochemistry at Texas A & M University; Dr. Virginia Swanson, professor of pathology at the University of Southern California; Thana Bisalputra, professor of botany and electron microscopist at the University of British Columbia; Dr. Abraham Rosenberg, professor of biochemistry at the School of Medicine at Pennsylvania State University; Dr. Gustav A. Paffenhöfer, of the Sylt Marine Station, Isle of Sylt, Germany, a postdoctoral research zoologist at Scripps; and Richard Lee, Scripps graduate student in marine biology.

Thus many different specialists are at work on many different aspects of the over-all copepod problem, and we can obviously do no more than sketch the main outlines of what they are doing and what they hope to learn.

Perhaps one of the first questions that one would ask would be why the copepods behave as they do. They have been hibernating in cold bottom waters, and then quite suddenly, when the spring diatom bloom occurs, they rise to the surface and feed voraciously on these one-celled algae. Do they receive some kind of signal that tells them that it is time to rise and feed?

A fascinating clue comes from the laboratory. There, James Kittredge, a student of Professor Denis Fox's, and Reuben Lasker, formerly of the Fishery-Oceanography Center of the U. S. Department of Commerce at La Jolla, raised copepods in captivity, and made many direct observations of their behavior as they pursued diatoms that the researchers put within their reach.

Repeated observations revealed a most astonishing kind of behavior. As Kittredge and Lasker watched an individual animal, they saw it dart in a straight line for a distance, and then stop and swim round and round in a tight circle. Soon it would dart in some quite different and apparently random direction, and then stop again and swim in a tight circle. It would repeat this pattern over and over, darts in many different directions followed by swimming in little circles.

Kittredge and Lasker believe that this is the copepod's normal search-and-seize procedure in catching diatoms. The dart is in a direction triggered by some biochemical signal, and the tight circling is a pause for the animal to get its bearings; that is to say, a hold pattern awaiting the next signal.

It was clearly established that this was not unusual behavior, but characteristic and typical, for in addition to their visual observations, Kittredge and Lasker made extensive permanent records in the form of holographic movies.[1] The question of interest then immediately became: What is the nature of this biochemical signal? Is it something akin to our sense of smell, or of taste? The current feeling is that it is probably analogous to smell.

What then might a copepod smell, and what would it use for a nose? The smell, of course, would arise from the prey—in this case the diatom. So the search began for a chemical compound that might be present in diatoms and that copepods might be able to detect in sea water. The sequence of investigations into the biochemistry and behavioral patterns that led to the current views is too complex for our consideration here, but the copepods are probably smelling polyunsaturated hydrocarbons

---

[1] Holography is another striking example of the power of modern research methods, and especially of how recently they have become available. It was not until 1965 that Emmet N. Leith and Juris Upatnieks of the University of Michigan used laser light to produce photographs that can be viewed in three dimensions. Thus Dennis Gabor's 1947 idea for holography had its first really useful application some twenty years later. Today, three-dimensional holographic movies of tiny subjects are of excellent quality. Even holograms in color are now possible.

that the diatoms produce. If we accept that as reasonable, we must next ask: What would a copepod use for a nose? And especially: How could it know what direction the odor came from, so that it could dart toward it?

At the bottom of the much enlarged photograph of a copepod that appears on page 268, you will note two antennae (pointing to the left and to the right) that look somewhat like the antennae of insects. Extending below the copepod's antennae are many fine hairs, and these, added to the main branches of the antennae, make up an arrangement somewhat like a complicated TV antenna. That, the Scripps scientists suggest, is what the copepod smells with.

But even granting that, how does he know what direction a diatom's odor comes from? Here is one of the most astonishing adaptations in nature. Most copepods have one red antennae and one white one. The function of this arrangement would be so that it could smell, as it were, in two directions. More—the color of the red antenna is caused by the same pigment (probably a carotinoid) that gives the salmon its red color. Since the life cycle of copepods is closely associated with that of baby salmon, this opens up a further fascinating line of investigation for some future researcher.

There is yet another mystery in the behavior of the participants in the copepod food chain. For as the copepods become plentiful when the diatoms become plentiful, so four-inch baby salmon come swimming down the streams where they were born and make the copepods their exclusive diet. How the baby salmon know when to do this is an even more elusive question than that of the copepod behavior, but it is again possible that a sense of smell may have some bearing on it; at least there is no question that salmon do have a sense of smell. After many years of research at their Wisconsin Lake Laboratory, Professors Arthur D. Hasler and James A. Larsen of the University of Wisconsin have conclusively shown that different streams in which salmon have hatched have different odors, that salmon respond to these odors, and that they remember odors to which they have been conditioned.

Although there may be some doubt about just how the baby salmon know that there are copepods to be had, there is no question at all as to why they hunt them. They do it because they are after the waxes that the copepods produce—waxes that are essential in the salmon diet. If you take a Vancouver copepod and dry him out under gentle heat, you will find that as much as half of his dry weight consists of stored liquid wax. This is the principal food of the baby salmon.

The salmon, then, is an ideal animal to use to study how these liquid waxes are metabolized. Through the most sensitive of analytical methods, Scripps scientists have found that the salmon fry first turn the liquid waxes into fatty alcohols, and then convert these into fatty acids. The significance in this finding is that fatty acids are precisely the "fats" that human beings find most digestible.

There is a further mysterious sidelight that enters into this process. Somewhere in the conversion of the liquid waxes from the copepods to fatty acids there occurs—as an intermediate stage—the production of fatty aldehydes. So far, none of these have been found in any salmon materials that have been studied, but Scripps scientists are eager to look for them in blood samples from young feeding salmon. Why? *Because fatty aldehydes are conspicuous components of human heart muscle and the nerves of the human brain,* and so far no one has explained why they should be there. Perhaps the answer will be found in salmon blood. It is a long jump from fatty aldehydes in the blood of salmon to fatty aldehydes in human organs, but when we consider the eons over which evolutionary changes from form to form have taken place, it may not be such a long jump after all.

The further the researchers pursue these matters, the more questions they ask and try to answer. Item: When copepods eat diatoms of one species, do they produce a different kind of wax from the kind they produce when they eat other species? Item: Copepods can survive long winters without eating anything. Obviously they do so by using stored waxes as a reserve supply of energy. This must mean that there is some way in which the rate of utilization of the energy reserve is regulated. The researchers believe that this is governed by enzymes, and they have good indications of what these enzymes are. What they want to know more about is how the enzymes work.

One of the laboratory methods that Dr. Benson and his colleagues will use in studying these questions will be to starve captive copepods and then feed them until they are completely recovered. From time to time throughout this process they will take samples of a few animals for analysis. In Dr. Benson's words:

"We suspect that the waxes of the [copepod's] fuel reserve, since they are less readily metabolized than normal fats, serve the animal in periods of stress and starvation. The waxes allow him to survive at a limited energy expenditure for longer periods than possible with normal fat metabolic control mechanisms. We believe our work will provide an answer to this question, and from the answer we should derive and define a multitude of ecological relationships among predator and prey in the marine environment.

"As our understanding of the distribution, metabolism, and transport physiology of wax esters in fishes grows, we will be in a position to apply these relationships to novel ecological systems. For example, it is possible that grazing and predation rates may be estimated from wax levels in the predator and the prey. The path of energy reserves in the food chain may be followed by the very characteristic and specific 'finger print' of dietary fatty alcohol and fatty acid chain lengths and desaturation. With a key to the relationships between fatty acids and copepod fatty alcohols, we will have a way to determine 'who eats who' in the ocean."

It should be noted that by "who eats who" Dr. Benson is not referring just to the diatom–copepod–salmon-fry food chain, but to many others that are not now nearly so well understood.

A few paragraphs back we noted rather casually that Dr. Benson is studying the metabolism of copepods by starving them and then feeding them back to a normal state of nutrition. That might suggest that raising copepods in captivity is a fairly simple matter, like raising the fruit flies that have been so widely used in genetic studies, or laboratory mice that thrive on a standard diet, or laboratory cockroaches that thrive on almost anything.

Raising copepods in captivity is a completely different matter. To begin with, there are many different stages between the fertilized egg and the adult. The egg hatches into a tiny, rather primitive larval stage called the nauplius, which has only three pairs of limbs. These limbs are at first used entirely for swimming, but as development progresses from stage to stage, they become more and more specialized. Furthermore, the nauplius is encased in a chitinous shell like that of a shrimp, which cannot expand, so the larva cannot grow gradually. Rather, it sheds its skin from time to time, each time becoming a little larger. In all there are six of these larval or nauplier growth stages, which are then followed by six stages in which the larva becomes more and more like the adult. The last six stages are called copepodite states, and as the animal matures from copepodite stage to copepodite stage, it adds more segments to its body until the adult stage is finally reached. Viewing prepared slides of each of these twelve stages in sequence under the microscope, one has a tremendous sense of seeing eons of evolution take place in what the animal has accomplished in a matter of months. The original pairs of primitive limbs have evolved into the specialized oar feet, into antennae, and, in the male of some species, into a kind of forceps for grasping the female.[2]

Thus from the standpoint of the animal's life cycle alone, the rearing of copepods for experimental purposes is a delicate business. A further difficulty arises from the copepod's dietary requirements. Again, this is no matter of setting out a feeder of prepared nutrient. Copepods require a diet of fresh live diatoms, so that a continuous support program of growing diatoms is also necessary. And, as in their home waters of the Strait of Georgia, they must be supplied with sea water containing the essential inorganic nutrients, together with the silica which they need to build their shells. Finally, copepods do not thrive unless they are fed in small amounts at frequent intervals, and Dr. Paffenhöfer and Professor Michael M. Mullen, assistant professor of marine biology at Scripps, have been most successful by feeding them pure cultures of diatoms on an hourly basis. By this tedious method they have succeeded in rearing captive copepods throughout their complete life cycles in the laboratory.

When the adult stage has at last been reached, the starvation and recovery experiments are ready to begin. To follow closely just what is happening to the metabolism of the copepod waxes, it is essential to take samples at frequent intervals throughout the entire sequence of experiments, and because of the difficulties of rearing captive copepods in the first place, it is feasible only to take very small numbers of animals at any one time. This means that only the most sensitive and delicate of analytical procedures will serve. This, as in much related work, calls for chromatographic methods.

You will recall the extremely sensitive method of paper chromatography that Melvin Calvin and his group used in analyzing the photosynthetic products of the one-celled *Chlorella*

---

[2] The male copepod has one of the most bizarre notions of how to perform the sex act to be found in nature. From a special gland he produces a plastic-like substance which he spins into a container that looks like a bottle with a very long neck. He discharges his sperm into this bottle, and then goes looking for a female. When he finds her, he fastens the neck of the bottle to an opening in one of her abdominal segments. The sperm enter a storage cavity in the female, ready to fertilize the eggs as they descend from the ovary.

plant, together with the checkerboard analogy that we used to help explain how it works. In the copepod experiments, the chromatographic principle was also used, but with the variation that instead of paper the Scripps scientists used thin-layer plates of silica gel, and in some cases chromatographic columns of silicic acid. And, as in Calvin's work, they used radiocarbon, C^{14}, so that the developed chromatograms could be exposed to medical X-ray film for heightened sensitivity. By this means they have been able to obtain accurate analyses of fats and waxes in samples consisting of no more than three copepods, each weighing no more than 1/140,000 of an ounce.

These experiments conclusively showed that the liquid waxes that copepods contain serve the animals, in this case *Calanus helgolandicus,* as a short-term metabolic fuel. More broadly, it also now seems clear that the liquid waxes that copepods manufacture are a major medium for storing and transferring energy in a variety of marine organisms throughout the food web of the ocean. Continued research along these lines is virtually certain to produce further revealing (and probably unanticipated) results.

As a notable example, who would guess that a study of the metabolism of liquid waxes in copepods of the Strait of Georgia would have anything to do with the low water requirements of animals of the desert? Yet there seems to be a close relationship between them.

According to Dr. Benson, the "classic" work of Dr. Knut Schmidt-Nielson of Duke University on the water requirements of desert animals explains how they adapt to arid conditions by developing a means of oxidizing fats that actually leads to the controlled production of water within their bodies. Similarly, whales and fishes, which cannot drink sea water, may be able to produce the fresh water that they need through the metabolism of fats or waxes. So we see that the phenomenon of fat and wax metabolism may be virtually as broad as the whole question of the accumulation, storage, and transfer of energy throughout living systems.

We know that in photosynthesis some plants produce fats rather than carbohydrates; that copepods convert these to liquid waxes; that salmon convert these waxes to fatty acids that are highly nutritious to humans; that some desert animals and probably some marine animals can convert fats into water; and that some of the intermediate products that occur during these metabolic processes are found in human heart muscle and in the nerves of the human brain.

The whole subject, however, is in such a state of rapid flux and development that at this point there can be no such thing as a summing up. That will have to be left to some future historian of science.

The point to be stressed today is that research in marine science, already well established but surely deserving of additional support, is surging forward at breath-taking speed, and that it promises such benefits toward the human good as to be truly beyond our imagination.

In discouraging contrast we hear from certain quarters that science is impersonal; that our technological age is robbing us of our freedoms and our individualism; in short, that we have nothing to look forward to but "big brother's" meddling and the tragedy of *1984*.

I find this attitude incomprehensible. The ocean scientists that I have met and talked with are not robots in white coats with their heads in the clouds. They are as human as the rest of us, and their overriding interests are in the welfare of mankind. They are working very hard to that end. Perhaps this chapter has been technical and difficult here and there; if so it is for that very reason that I would cite its conclusions as further evidence for man's eventual understanding of the ocean world—and his disposition toward the wise use of it.

20.
CONSERVING THE BOUNTY OF THE SEA

One of the widely disputed questions of our day is how much the food resources of the ocean can contribute to the food needs of an exploding population. The subject is obviously of critical importance, but before we go into it, let us ask if this is really the right question.

In the first place, it is only one part of the much larger question: Can the world's total capacity for producing food support an exploding population? We know a great deal more about farming the land than we do about farming the sea, so it seems logical to look first at what we may expect from the land. Knowing that, we will be in a better position to ask what we should expect from the sea. The land has always produced far more food than the sea, and there seems no reason to suppose that this will change. Quite the contrary, for recent research suggests that the land may come to have an even more dominant role.

In earlier chapters and other contexts, we noted two developments of outstanding importance. First, Dr. C. T. de Wit of the Netherlands made the highly sophisticated calculation that the photosynthetic capacity of land plants could support a human population—explode as it may at the present exponential rate—for hundreds of years to come. To put actual food on the table from this potential source involves political, social, and economic problems that Dr. de Wit did not go into, but his argument that the potential is there to be tapped seems convincing. A critical point that he failed to mention is that the world food problem lies more in a shortage of protein than of carbohydrate. But other research, in an entirely different area, suggests an answer to that also.

You will recall that after Melvin Calvin had elucidated the path of carbon in photosynthesis, two of his associates, James Bassham and Richard Jensen, saw promising evidence that chemical controls may make it possible to increase the amounts of proteins and fats that green leaves produce. This would hardly suggest that most of us would care for such leaves dried and packaged as breakfast food. But it does suggest that such protein-rich leaves would be an excellent animal feed that could relieve much of the pressure that we are now putting on our fisheries.

According to S. J. Holt, a fisheries authority with the Food and Agricultural Organization of the United Nations, the Peruvian anchoveta is caught in the greatest annual tonnages of any fish in the world. It is not used for human food, but is processed into fish meal as an animal food supplement. Indeed, we are told that if it were not for cheap fish meal, the housewife would find chickens one of her most expensive protein sources rather than one of the cheapest. Surely it would be cheaper to harvest protein-rich leaves from tropical land vegetation than to hunt anchovetas off the coast of Peru. Of course the potential of tropical farming is not proven, but this very recent finding has extraordinary potential, and it is no fantasy to see a bright future for its benefit to mankind.

The whole history of science and technology shows that a highly efficient way to go wrong is to base our expectations on the "foreseeable future." We could cite case after case, but just one, concerning the world's energy supply, will make the point.

Not long after the end of World War I we became greatly alarmed about the rising rate at which our burgeoning industries were depleting the world's store of energy sources, chiefly of coal and oil. Many estimates were made (and stamped with official authority) of how soon the oil wells would dry up, how expensive it would be to go deeper for coal, and how frustrating were the technological problems of obtaining energy from oil shales. It therefore became imperative to consider other possible sources of energy.

It was well known that as early as 1905 Albert Einstein had shown that an incredible store of energy was locked within the atom. Why not tap that? Was it possible? What was needed was a study of the technical feasibility of releasing and using energy from that source. Such studies were made, and as usual they were carefully based on the "foreseeable future."

The results of one of them were published in 1928, in a book by Thomas William Jones called *Hermes, or the Future of Chemistry*. It seems to have been well received at the time, for H. M. Parshley, reviewing the book for the *New York Herald Tribune,* said, "This latest little volume in the series makes a serious straightforward study of its theme. . . . The reader gets from this volume dependable information on the present status of chemical knowledge."

Some of this "dependable information" concerned the future of energy from the atom, and of that Jones wrote:

"The sensationalist press and the novelist of the fantastic have led the public to expect early and far-reaching developments through the application of what is vaguely termed 'atomic energy.' But the cost of artificially disrupting so stable an electrical system as the atom does not bring it within the range of practical usefulness. Experiments show that, on the average, the cost is something like 100,000 times the gain, and there is no evidence that the cost can be decreased to an economic figure. It has been calculated that one gram of radium would have to project its rays into a sheet of aluminum for a period of 5,000 years to disrupt the aluminum atoms sufficiently to yield a cubic millimeter of hydrogen—scarcely an explosive reaction. . . . It is difficult to see wherein lie the advantages of this overrated atomic energy. The most we can say at present is that it *may* turn out to be fresh material for us and it *may* be harnessed, but there is not at present the slightest indication that it will, nor that our traffic problems of the future will be resolved by any other energy than that derived from the resources of water and fuel."

So wrote Thomas Jones in 1928. In 1932, James Chadwick discovered the neutron, and physicists soon realized that the neutron was the key to unlocking the nucleus of the atom. In the laboratory at least, the release of nuclear energy was not far off. Working in Italy, Enrico Fermi produced radioactive isotopes of more than thirty different elements. What followed in the next three decades is common knowledge.

This is not to denigrate Thomas Jones. But it is to question the general credibility of forecasts based on the "foreseeable future." The future of an abundant production of plant protein by altering the course of the photosynthetic reaction in green leaves is admittedly speculative now, but it would surely be surprising if that—or some unforeseeable alternative—did not emerge in the next few years.

This brings us back to the question that we started with. How much can the ocean contribute to the food needs of an exploding population? Given now the promise of an abundant supply of protein from the land, we must ask—even more insistently—is that the right question? Probably not. A better question would seem to be: How can we manage our ocean resources so as to reap the most rewarding harvest in terms of quality rather than quantity? Given that freedom of choice, we would look for two benefits: (1) to meet nutritional needs, a good yield of high-quality animal protein; and (2) to meet culinary preferences, a really appetizing product. As a matter of choice, "bread" from algal flour is unlikely to replace grilled Dover sole.

In stating the desirability of an appetizing product I do not mean to imply any lack of concern for the world's millions that are starving for protein of any kind. Their plight is tragic and humanly indefensible. There is an appalling difference in the way that people's food needs are treated from one area of the world to the next. Throughout North America the average person gets some two and a half ounces of animal protein a day. In the Far East, the average person gets less than a quarter of an ounce. This means that some get none. The average man in the Far East gets perhaps a third more vegetable protein than his North American counterpart, but that is not enough to keep him from being hungry every day of his life.

For our part, we must surely support the efforts that are being made to help, especially at the international level. For his contribution to the "Green Revolution," Norman E. Borlaug received the Nobel Peace Prize in 1970, and he has continued to develop improved methods of producing wheat in deprived areas. The Food and Agricultural Organization of the United Nations is working hard on better ways to get food from both land and sea, and with respect to the sea, for transporting the product for long distances inland. Other problems being faced are how consumers can pay, how primitive fishing methods can be improved, and how basic knowledge about fisheries biology can be made more widely available. And although it will be small comfort to deprived areas, it is clear that even the most highly developed areas still have much to learn about the wise management of ocean resources.

Not long ago Dr. William A. Nierenberg, director of the Scripps Institution of Oceanography, told Don Dedera, "The richness of the American way of life for centuries to come may depend on how wisely we *explore and use* the oceans in the coming decade. The 1970's, which history may mark as the Decade of the Deep, could open a new age of scientific discovery and economic benefit surpassing the returns of the just-ended decade of space exploration."

In what directions is this exploration likely to be most fruitful? As guidance here we are fortunate to have available one of the most thoughtful studies that has been done in our time. It was carried out under the sponsorship of the Conservation Foundation, which was established in 1948, and which by its own statement "is an independent organization established to promote greater knowledge about the earth's resources—its waters, soils, minerals, plant and animal life; to initiate research and education concerning these resources and their relation to each other; to ascertain the most effective methods of making them available and useful to people; to assess population trends and their effect upon environment; finally, to encourage human conduct to sustain and enrich life on earth."

The man who carried out the study, with the help and cooperation of experts from all over the world, was Dr. Lionel A. Walford, who at the time was chief, Branch of Fishery Biology, of the U. S. Fish and Wildlife Service. An authority on marine ecology and biogeography, he had earned his A.B. at Stanford and his Ph.D. in biology at Harvard. He taught marine biology in several universities and served in various capacities of increasing responsibility in the U. S. Fish and Wildlife Service. He was thus eminently qualified to carry out the study that the Conservation Foundation felt an urgent need for, and which was ultimately published in 1958 under the title *Living Resources of the Sea: Opportunities for Research and Expansion.*

Dr. Walford set forth the basic problem that he explored in his study in the following way:[1]

> Some time ago, The Conservation Foundation, an organization devoted to the proper use of natural resources for the welfare of mankind, asked me to explore the following question:
>
> *What scientific researches, apart from those which are in progress, would contribute significantly toward learning how to enlarge the yield of food from the sea in answer to human needs?*
>
> Back of this question are these assumptions: (1) We have not yet learned how to exploit the food resources of the sea fully; (2) scientific research will show the way; (3) there are gaps in present research programs that need filling. What are those gaps? [We] begin by discussing the world food problem. Populations in various parts of the world are suffering from ills caused by protein deficiency. Much more protein food of animal origin is needed than is now being produced, and still more will be needed in the future. The sea evidently does have untapped food resources, but we do not know how extensive they are, nor how to exploit some of them; nor have we explored fully all of their possible uses. This ignorance limits exploitation of the sea's food resources but so also do factors in the field of economics and sociology.
>
> A great deal of research about the sea and its resources is going on in many parts of the world. Most of this is conducted by governments, and for the most part it is concerned with established fisheries and is aimed directly at practical application. Applied sea fishery research is given considerable attention. It is in the realm of pure or fundamental research that the most important gaps occur and where augmented support is most needed. This book will explore these gaps.

I shall not now attempt any condensation or review of Dr. Walford's *Living Resources of the Sea*. The book is a milestone, and anyone who is concerned about the food resources of the ocean should read it. What does seem well worth doing here is

---

[1] Lionel A. Walford: *Living Resources of the Sea: Opportunities for Research and Expansion.* Copyright © 1958. The Conservation Foundation; The Ronald Press Company, New York.

to add some notes and comments on some of the research results that have been obtained since the book was written.

You will have noted, for example, that in stating his problem, Dr. Walford made the flat statement that "more protein food of *animal origin*[2] is needed than is now being produced." That, of course was written before Bassham and Jensen saw a possible way of increasing the production of protein by plants. Other notable advances have been made since Walford wrote, and the following five sections take note of some outstanding ones among them. This in no sense is intended to bring the whole vast subject up to date. It is simply to indicate a number of encouraging signs, with the object of giving some current support to Dr. Walford's basic conclusion that our greatest need is for more support of basic, or fundamental, research in the marine sciences. If more support was needed in 1958, it is needed even more urgently now.

CONSERVATION

Next to "ecology," "conservation" may be one of the most misused terms of our time. There is no need here to go into the emotional outbursts to which the press and the electronic media expose us in the name of conservation. Rather let us turn to the dictionary, which defines conservation simply as "preservation from loss, injury, decay, or waste." In our present context, Dr. Walford calls it "the intelligent use of resources," and goes on to develop that idea by saying that the ideal of scientific conservation is to permit a maximum sustained productivity of an ocean resource, involving "full utilization for the benefit of mankind; not restriction for the benefit of fish."

Certainly one of the most serious threats to this ideal of scientific conservation is the enormous amount of public misinformation that exists about it. Much of this misinformation belongs in the category of folklore; and two of the most unfortunate aspects of it are that it is well-intentioned and that it is based on what seems to be common sense. The opinions "stand to reason," and that is their insidious danger.

When John Steinbeck was writing *Cannery Row* and Ed Ricketts was studying the ecology of tide-pool animals at Pacific Grove, the commercial catch of Pacific sardines was just past its all-time peak. It had begun with a small commercial cannery at

---

[2] Emphasis mine (R.W.).

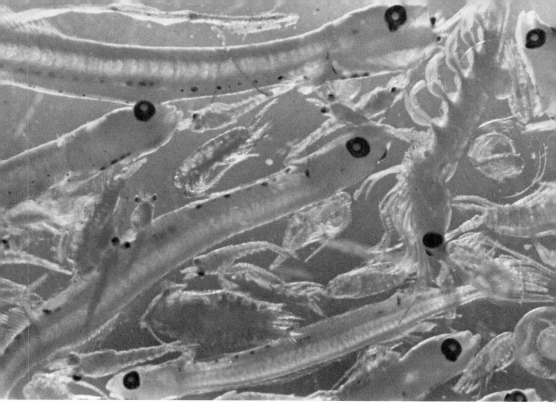

This sample of plankton from the California Current contains anchovy larvae, copepods, sardine eggs, and, to the right, an euphausiid.

North Beach on San Francisco Bay in 1889, and it grew to be the largest fishery in America. In 1936, the catch was one and a half billion pounds.

Since then, the trend has been steadily and sharply downward. By 1951 fishing out of San Francisco ceased, and by 1960 the Monterey canneries were out of business. The fish were gone.

What caused this catastrophic decline in a great resource? The common-sense answer, of course, was overfishing. The California legislators certainly thought so, for they passed laws regulating the proportion of the catch that must be canned for human food in relation to the millions of pounds that were being processed into fish meal for chicken feed. But to get around that, some enterprising entrepreneurs went outside the three-mile limit with factory vessels equipped to make meal on board; and an even louder cry went up that if overfishing were not stopped the fishery would die. It did die—so what more proof was needed for the popular mind? A more critical mind would have insisted that overfishing was an oversimplified explanation, and that the failure to maintain a maximum sustained productivity may have had far deeper causes.

I must confess that I fell into this oversimplification trap myself. One of the most thrilling experiences in my memory was aboard a sardine purse-seiner out of San Francisco at about the time that the sardines were most plentiful. It was a memorable experience to hear the lookout cry "Fish, fish!" when in the dark of a moonless night he spotted a great school of sardines from the phosphorescent glow in the water. To see the net of the purse-seine paid out, and watch the boat circle the school, and hear the care with which the captain gave orders for the pursing of the net as though by a drawstring—carefully because a slight disturbance might cause the whole school to sound and escape. Then the long task of hauling the net, and shoveling ton after ton of silver fish into the hold, and when the job was at last done, to relax over a great hunk of Italian bread and a tumbler of red wine while the boat turned and headed back for the cannery.

So, being emotionally involved when the sardine fishery began to decline, I was quite ready to accept the idea that overfishing was the whole cause of the decline. I could not have been more wrong, and in accepting the conclusion I assumed that I knew the answer. I did not.

If overfishing was not the only cause of the demise of the fishery, what were the other causes? The painful answer is that no one can be sure. If all the causes had been known at the time, could the fishery have been saved? We do not know that either.

Much too long after the fact, marine biologists began looking for other causes, and they found some very reasonable possibilities. One of these is that some change in the environment caused a sharp decline in the rate at which young fish were added to the population. This may have been due to a declining rate of reproduction, or to a declining rate of survival of fingerlings. But whether it was one or the other or both, it continued for a number of years in succession. Therefore, for a long time replacement of the population took place at an abnormally slow rate. During that period (if it had been recognized for what it was) wise exploitation of the resource would have called for a reduction in the numbers of adult fish taken. One cannot make continuous withdrawals from a bank account without making some deposits.

Another possibility that was suggested was that the northern anchovies, which invaded the area when the sardines began to decline, began to eat all the food that the sardines depended on. In that case, the sardine population could only decline further or go somewhere else in search of food.

These are the principal reasons that have been suggested, but there may have been others as well. To learn more about these it would have been necessary to carry out *continuing* studies such as those that Dr. Walford recommended for the scientific conservation of any fishery. Answers to questions like the following would have been essential.

- How were the sardines distributed geographically throughout the entire year, including the outer limits of the productive fishing areas?
- What was the true abundance of the sardine population?
- What was the average growth rate of individual fishes?
- Dividing the sardine population into various age groups, what were the habits of migration of these groups?
- What were the spawning habits? When was the chief season and how long did it last? How many fishes survived to various ages per million eggs deposited?

The northern anchovy. While it supports a commercial fishery of some 100,000 tons per year (and is a favorite bait for sports fishermen), it may have been partly to blame for the demise of a sardine fishery seven or eight times as large.

• What was the relationship between the rates at which the sardines were taken and the real productivity of the fishery?
• What were the mechanisms by which fluctuations in the environment affected the natural mortality of fishes of various ages?
• What features of the environment most affected the distribution and other habits of the sardine population?

But answers to these questions were not sought. If they had been, could the fishery have been saved? There is no way of knowing. That there was mismanagement of the resource is obvious enough. But would even the best management have saved it? Again, we do not know. That is the essence of the tragedy. It is not so much that the fishery died. The real tragedy is that we might have known why it did, and we did not try in time.

Well, it is too late for the California sardine fishery, but it is not too late for others, and society should demand that the best available methods of scientific conservation be followed in the future. To do this, however, much more knowledge about the ecology of the seas than we now possess is essential. The first area in which more knowledge is needed is that of the marine environment itself.

THE MARINE ENVIRONMENT

The moment we start talking about the problems of the environment we run into a morass of misinformation and misunderstanding. There has been as much "common sense" nonsense bandied about the environment as there has been about conservation.

We may illustrate this with an example that is not directly related to marine problems but shows especially well the contributions that basic research can make to practical issues. It too is a product of very recent research. The American Chemical Society cited it as one of the most important chemical developments in 1972.

It concerns the occurrence of carbon monoxide in the atmosphere. We know that although it is a colorless and odorless gas, carbon monoxide is a deadly poison, that it has a role in the production of irritants in smog, and that it results from the incomplete combustion of fuel. It is present in automobile exhausts, in smokestack gases, in the effluents from inefficient domestic heaters, even from open burning in areas where that is still permitted. Industrial man, therefore, with his huge consumption of wood, coal, oil, and gas, and the partial combustion that accompanies much of it, must be releasing enough carbon monoxide into the atmosphere to endanger the ecological balance. That is ordinary common sense.

Early in 1973, however, Thomas H. Maugh II reported in the journal *Science* on a 1972 development that the American Chemical Society had singled out. Working at the Argonne National Laboratory in Argonne, Illinois, C. M. Stevens asked this question: "How much carbon monoxide is released into the air from man's activities, and how much of it comes from natural sources?" Being at Argonne, Stevens had ready access to advanced methods of measuring the ratios of the carbon and oxygen isotopes in the carbon monoxide that man produces and that which comes from natural sources. What he found was that each year man in the Northern Hemisphere produces less than three million tons of carbon monoxide, and nature produces more than three *billion* tons, the latter by oxidation in the atmosphere of methane from decaying vegetable matter. Of course it makes good sense to try to reduce the three million tons that *we* produce, but even if we do not, we will be doing little to disturb nature's own ecological balance.

That brings us to another example, this time not of folklore versus scientific evidence, but of more modern scientific views versus older "classical" ones.

Recall from an earlier chapter that every spring there is a great bloom of plant plankton (diatoms and flagellates) in the Strait of Georgia, and this bloom attracts countless copepods that are vital to the energy relations of the sea. The first link in this food chain is the population of photosynthetic diatoms and flagellates on which the copepods graze, and the classical view has been that the tiny plants bloom because upwelling carries a rich supply of mineral nutrients to the upper waters that sunlight can penetrate.

S. J. Holt, however, has recently called attention to the view that the plants may not be seeking the mineral nutrients so much as the cooler environment of the upwelling waters. Offhand, this seems like a fine point, but it may have a thoroughly practical application. Consider the oversimplified picture of the fisherman looking for schools of fish that are looking for copepods that are looking for diatoms. The fisherman will be more interested in small areas where the fish are highly concentrated than in broad areas, even though such broad areas are alive with fish. It may well be more helpful to the fisherman if he has reliable information on where the cool water areas are most likely to be concentrated from year to year. From this example too, we see again the vital importance of *continuous* observation of the conditions of the environment. That brings us to a consideration of the endless variety of creatures that are found in the environment.

THE IDENTITY OF SPECIES

For at least half a century after Darwin published *The Origin of Species,* biologists devoted a major share of their time and attention to the identification of species of plants and animals, and in discerning the relationships among them based on evolutionary principles. The science of identification of species was called taxonomy, and it was mostly carried out by specialists who had access to great collections. The fifty volumes of *Challenger* Reports represent a great study in the identification of species, and of course it was of extraordinary value. But taxonomy can be tiresome work, as Alexander Agassiz well realized when he had finished his study of the echini from the *Challenger* collection.

Aside from that, all kinds of exciting new fields of study in biology were opening up: physiology, genetics, and behavior, for example, which involved laboratory experiments and field studies that seemed far more interesting than classifying museum specimens preserved in bottles of alcohol. So taxonomy fell into a kind of limbo from which it did not recover for a long time.

But it could not die, for every biologist, whether he was studying the physiology of bioluminescent fireworms, the genetics of oysters, or the homing behavior of salmon, had to know exactly what species he was dealing with. Ed Ricketts could not have known that the specimens that he and John Steinbeck brought back from the Sea of Cortez included some 550 different species, of which fifty or so were new to science, unless he had had the benefit of taxonomic information and advice from museums and libraries in Mexico and California.

Fortunately, then, taxonomy has recently taken on a new and far more interesting aspect. The evolution of Charles Darwin has undergone some startling changes, and a modern theory called synthetic evolution has emerged. Based on new principles of genetics and new findings of molecular biology, it gives a whole new view of the diversity and especially the unity of life on earth. This has been brilliantly set forth by Ernst Mayr in his recent book *Populations, Species, and Evolution*. And as taxonomy sprang from Darwin, so a new discipline of the speciation of plants and animals has emerged from the theory of synthetic evolution. Of course, it had to have a new name too: "the new systematics."

It is an exciting study, but unfortunately it has not attracted a great deal of support. Whether he wishes to be known as doing taxonomy or as doing the new systematics, the specialist who identifies species still has to have specimens to work with. This means a great collection of some sort, whether it happens to be housed in a musty museum or on neat steel shelves in a modern laboratory of marine biology. Neither of these are receiving financial support in proportion to the contributions that they could make.

Not long ago I was shown through the collection of specimens of one of the world's great marine laboratories by an enthusiastic guide, and he pointed to thousands of specimen bottles —including some invaluable ones—that sat in disarray unstudied and unlabeled on the floor. There was not even decent storage space available for them, let alone the time and talent of specialists in the new systematics to put them in order. It was as though the Library of Congress in Washington were to receive some newly discovered Mark Twain manuscripts and shoved them into a dusty basement corner.

Why should this be so? There are a number of reasons. For one thing, there are fashions in science, and what is popular today may be quite out of style tomorrow. Then, too, a significant fraction of the money that supports scientific research comes from government grants, and those who hold the pursestrings on grants tend to favor work that promises to show an immediate practical benefit. Other causes would involve apathy, indifference, misunderstanding, and sheer ignorance.

The remedy, if one is to be found, will lie in the general recognition that science is a unified whole and cannot be expected to function well if some of its organs are ill-nourished. Our attack, to put it in the painfully obvious terms that this case seems to require, should be on all fronts. That is why it is not such a long jump to move now from systematics to the behavior of the animals that it describes.

ANIMAL BEHAVIOR

In the study that we mentioned earlier, S. J. Holt, who is director of the Division of Fishery Resources and Exploitation in the Department of Fisheries of the Food and Agricultural Organization of the United Nations, has estimated that the world catch of fish could be increased about threefold if the best methods of improving catches were to be generally observed. For the United States, he estimates that the present annual catch could be increased from some six million tons to about twenty million tons. Here he is assuming that improved methods of all sorts will be used. And it is to be stressed that no more fish would be taken than sound conservation practices would allow.

Some of the improvement will come from expected refinement of present methods, but most of it is likely to come from quite radical departures from tradition. As has happened again and again, these are most likely to result from unexpected new knowledge—sometimes acquired by chance observation, but most often through basic research.

The study of animal behavior is one of the newer (and fortunately one of the more popular) fields of study in biology today; and it is obvious that the more we can learn about how marine animals behave, and more especially why they do so, the closer we are likely to come to real breakthroughs in improved fishing methods. Let us look at a case in point that Dr. Walford has described.

"Fishermen locate schools with the echo sounder, but scientists, whose skill it is to arrange and collate data so as to bring out otherwise obscure patterns, can make much more from the records than just that. One example of a fishery biologist's studies of bathygrams will suffice to illustrate the point:

"By continuous use of the echo sounder, I. D. Richardson, of the British Ministry of Agriculture and Fisheries, was able to keep a research vessel over a school of sprats in the Thames Estuary from mid-afternoon of one day until mid-morning of the next. In the afternoon, the school was in shallow water, packed in a dense mass close to the bottom. About an hour after sunset, it started to rise. By 4:55 p.m. it had reached eighteen feet from the surface, and in the next twenty minutes as the light of the sun left the sky, it rose until the noise of the fish breaking the water could be heard. The school moved about, into areas that were deeper, but stayed near the surface until dawn, when it descended again towards the sea floor.

"He went on further to observe the diurnal movements of herring schools off North Shields, off the Yorkshire coast, off East Anglia, and off Cape Gris-Nez on the French coast. Herring evidently behave differently in different places. . . . When [Richardson] statistically analyzed the tracings from all places where echo soundings had recorded herring, it transpired that the level which schools take is determined by the intensity of light; and that varies from place to place and from time to time, depending on the quality of daylight and the turbidity of the water. A rapid swim certainly does occur at East Anglia, for fishermen's nets do on occasion fill up suddenly and fast, but it has yet to be explained."

This work with the echo sounder was a long stride ahead of the work that young Alister Hardy did in the early 1930's with his various versions of the plankton recorder as an aid to herring fishermen. But what we are really looking for is not the long stride but the great leap. That great leap seems now to be in the making. It will not rely on sound traveling through the water, but rather on electromagnetic radiation traveling through space.

In *Our Changing Fisheries,* a publication of the U. S. National Marine Fisheries Service, Frank J. Hester of the Bureau of Commercial Fisheries discusses some new and very promising uses of electromagnetic radiation of various wavelengths for remote sensing devices. Through these it will be possible to measure from considerable distances physical, chemical, and biological properties of great stretches of the ocean.

> These sensors [Hester wrote] can be mounted in aircraft and spacecraft, and will provide rapid coverage of huge areas of the ocean's surface. Already in use are infrared sensors for measuring surface water temperatures and detecting the passage of fish schools. Infrared temperature measuring sensors—combined with still-experimental radars that can measure wave heights, ocean currents, and changes in salinity—may one day allow oceanographers to learn in hours from spacecraft what would take months to learn from a ship.
>
> The visible portions of the electromagnetic spectrum have great promise too. Most areas of the ocean contain microscopic organisms which, when disturbed, give off flashes of light—known as bioluminescence. At night fish schools

disturb these organisms and create glowing balls of light which guide fishermen to their catch. Now through the use of electronic light intensifiers, much more sensitive than the human eye, bioluminescence can be detected from high-flying aircraft and perhaps spacecraft. Such measurements can be used to map the abundance of aquatic resources and to direct fishermen to the fish.

Improvements in the human visual sightings of fish also have been made for daytime use. With advanced techniques in color separation—either through photographs or with electronics—it is possible to enhance extremely slight color differences in the sea. These techniques bring out and identify fish schools, areas of pollution, changes in the amount of plant life in the water, and changes in water depth, all of which would not be visible to the human observer.

Ultraviolet sensors, operating beyond the blue part of the visible spectrum, can detect changes in the surface film of the ocean caused by the presence of animal and plant oils. These oils disappear quickly when exposed to air. Their presence on the sea surface provides clues to the location and identity of subsurface fish schools.

As a member of the staff of the Bureau of Commercial Fisheries, Hester is understandably concerned with improved methods of catching fish. But it takes little imagination to see the usefulness that these sensing devices would have in support of the basic research that we have been primarily concerned with. In either application, these are tools for the open ocean. Let us turn now to the tidal waters near the shore.

FARMING TIDAL WATERS

From time to time we read news accounts of enterprising people who operate aquatic farms in brackish inshore waters of bays and estuaries. They make the news columns because they are unusual. Dr. Walford has said that opportunities like these are completely neglected in most countries in spite of their great promise. Indeed, Walford says that inshore waters can produce larger quantities of animal protein than farmland can. At the same time, "our use of the sea as a source of food and other biological raw materials is technologically and philosophically about 200 years behind our use of the land."

Why should we put up with such an antiquated approach? Perhaps we have not consciously put up with it, but have simply drifted into it—largely through lack of information about the potential that lies at the water's edge.

This is not to imply that you can throw oyster spat into the water at Oxford, Maryland, as a corn-belt farmer sows his seed, and expect to reap a bumper crop a few months later. There are all kinds of difficulties in the way of oyster farming, and yet it is done commercially, and very successfully too. For example, even some time ago the Johnson Oyster Company near Drake's Bay in California was supplying the local market with more than four million oysters a year. The standing crop in that sheltered lagoon is worth more than a million dollars.

The tiny Olympia oyster, the gourmet's delight for its so-called "coppery" flavor, has been successfully farmed in the state of Washington for many years.

"Have you heard the story about the farmer who planted ten acres of abalone?" asks *P. G. & E. Progress*. "If this sounds like the start of a tall tale, meet three men who are doing just that: John Perkins, Hugh Staton and Dr. David Leighton.

"On a coastal highland in San Luis Obispo County, they are growing a crop of about a million abalone—something that most experts had predicted couldn't be done. In fact Staton himself, when majoring in zoology at the University of Southern California, had written a paper stating that it was not feasible.

"Yet the trio have completed three years of a pilot operation and are now setting out to triple their spawning facilities—adding a big new pond and another building. Their apparent success has drawn world-wide attention, and their goal of eventually marketing a million abalone a year now seems possible."

Leighton is a marine biologist, and "the three men soon absorbed the slim body of knowledge (mostly from Japanese sources) that exists on the subject, and went on to develop new information . . . in the laboratory-nursery they built near Cayucos."

We may conclude this section with mention of a very recent study of oyster culture. Working at the Biological Laboratory of the Bureau of Commercial Fisheries at Milford, Connecticut, A. Crosby Longwell and S. S. Styles have been studying the genetics of the commercial American oyster, and have come to the conclusion that the regular introduction of wild strains is essential to the continued success of artificial commercial beds: this, as it were, to prevent inbreeding of the "domesticated" stock. They suggest that for optimum success in commercial oyster growing a greater understanding of the genetics of the oyster is needed, together with the application of that information in ways like those that are used in breeding better varieties of plants. Once more we are impressed with how useful basic research may be.

The foregoing discussions have been so wide-ranging that it might seem next to impossible to bring them together into any kind of intellectually manageable whole. Yet advanced techniques for doing just that, and more, are being developed, as we shall now see.

WEAVING THE STRANDS TOGETHER

We come now to one of the most advanced concepts in modern science. It involves studying the whole natural world, or any part of it however small; or the entire universe, or any part of *it* however small. The application of this concept is called systems analysis, and it includes any means of collecting information about any facet of the real world, and reducing that information to some useful form that the human mind can grasp. Very often the amount of information to be processed is so large that only the most advanced computers can handle it.

First, what is a system? The dictionary says that a system is "an assemblage of or combination of things or parts forming a complex or unitary whole, as, a mountain system or a railroad system." Dr. D. B. Luten, who has thought for a long time about systems guidance problems, says that a system may exist as much in the mind as it does in nature, and defines a system as "an object of interest together with its significant environment."

We may regard any system as a component of some larger system (called a supersystem), and itself as made up of a number of smaller systems (called subsystems). In biology, a biome is a system, for a biome is defined as a large, easily differentiated community unit arising as a result of complex interactions of climate and other physical and biological factors. In Dr. Luten's terms, the community would be the "object of interest," and the rest would be "its significant environment."

Naturally any *marine* biome would be a system. Ed Ricketts's habitat areas of protected outer coast, open coast, bay and estuary, and wharf piling would constitute systems. The Strait of Georgia would be a system in which the copepod community might be the object of primary interest. What is most important to emphasize is that to really understand it, we must study the *whole* system.

gains must be put to use for the benefit of mankind, and seeing to that is not the primary business of the scientist.

It is the business, primary or otherwise, of all of us, including the concerned citizen, the teacher, the research director, the administrator, the entrepreneur, and the politician. In making the best use of scientific advances, we face social, political, and economic problems of great difficulty; but in spite of that, impressive progress has been made ever since Bacon, and is continuing to be made today.

It is encouraging to see dedicated men fight to achieve these goals, and this book will close with some notes about how a few of them have gone about it. These will undoubtedly call other examples to your mind, and perhaps suggest ways by which our collective efforts might be improved.

THE CONCERNED CITIZEN. I am fortunate in having as a personal friend one well-known concerned citizen whose efforts locally toward preserving the environment have been remarkably effective. He happens to be highly educated and broadly experienced, and that helps. But that is not the real key to his effectiveness. The key is that he is willing to spend long hours outside his regular work in learning everything he can about the subjects of his concern.

One of these subjects happens to be the effect on the environment of nuclear power plants. In considering what his position should be, he does not jump to the conclusion that nuclear plants should be banned because of their potential hazards—explosion, escape of radioactive materials into the atmosphere, or damage to river or ocean environments by heat. Rather, he studies how real these hazards may be, and tries to balance them against our inescapable need for more power than our available natural fuels and hydrogenerators can produce. His conclusions—which he is quite willing to modify in the light of new facts—have always seemed to me to make good sense.

He is articulate and entertaining, and although he enjoys talking about sports and politics and the merits of certain Bavarian beers, you are not likely to leave a lunch with him without some new thought about the environment. He also takes time to write letters to the editor and to his representatives in government. He is a most effective concerned citizen.

There is no reason why any intelligent person who is especially concerned about the marine environment should not follow his example; and naturally I hope that this book, and especially the suggested readings that follow, will heighten such interest.

THE TEACHER. It is quite impossible for me to think about inspired teaching without thinking of my friend Joel W. Hedgpeth. He has made teaching his life work, and he points to his students as the proudest product of his efforts.

Of course, teaching is done in many ways. It can be done through the writing of books, through classroom instruction, through demonstrations in the field, and, perhaps most important, by example. Joel Hedgpeth has done all of these.

After he received his Ph.D. in biology from the University of California, he joined the staff of Scripps Institution of Oceanography, where one of his chief responsibilities was the compilation and editing of the now classic *Treatise on Marine Ecology and Paleoecology,* which was published by the Geological Society of America in 1957. Volume I, on marine ecology, covers its subject in the thorough detail that one would expect of a book of 1,296 pages. Its many chapters are by world authorities, and in his capacity as editor and member of the National Research Council Committee on the project, Hedgpeth visited many of the authors here and abroad for discussions of the subject matter and the best ways of presenting it. He wrote the introduction himself—an invaluable survey of the scope and history of marine ecology—together with a number of the chapters on particular subjects.

This is his longest work, but perhaps not the most widely influential. He also assumed the responsibility for keeping *Between Pacific Tides* up to date. But probably the most widely used is his little paperback called *Seashore Life of the San Francisco Bay Region and the Coast of Northern California.* It was first published by the University of California Press in 1962 and went into its fourth printing in 1970. It can be found in any library and for sale at any seaside settlement along the central California coast. How many tide-pool novices and veterans alike have carried it with them to the shore would be impossible to guess.

There is an infectious enthusiasm about Hedgpeth's writing that makes it irresistible, and that—added to his accuracy of fact and clarity of expression—makes his work a most effective contribution to learning.

After the *Treatise* was published, Hedgpeth accepted an appointment as professor of zoology and director of the Pacific Marine Station of the University of the Pacific at Dillon Beach. That is a short sail from San Francisco Bay, and that is where he taught the students who are now contributing effectively to solutions to the many environmental problems of the Bay.

In 1965, Dr. Hedgpeth accepted his present post as professor of oceanography and director of Oregon State University's Marine Science Center at Newport. From Newport as an operating base, he roams the world, carrying the message of environmental quality to audiences at all levels. He can be found teaching a seashore class in marine ecology at Puerto Penasco at the upper reach of the Sea of Cortez, addressing a chamber of commerce on the history of oceanography, teaching a university extension class about the San Francisco Bay environment, consulting with nuclear scientists at Oak Ridge, advising state and federal bodies about problems of estuaries, or addressing an international audience in Edinburgh, Scotland, on "the flowering of seashore books."

His platform style is as effective as his writing, for he prepares with a flair. Late one Sunday afternoon he stopped by our house and told us that he had spent the afternoon prowling the antique shops in Port Costa in search of an antique saber. He was lecturing the next day, and he wanted to wave an old saber to illustrate one of his points.

Finally, an inspired teacher sets an example. In addition to all of his other qualities, Hedgpeth sets a redoubtable example for plain hard work. He will accept any obligation at any time to forward the cause of conservation. He is what, in the best sense, Galsworthy could have called a "man of influence."

THE RESEARCH DIRECTOR. The research director of the kind that I refer to here does not exist in marine science yet, but I suggest that the establishment and support of such positions would be most rewarding to the ocean industries that depend so heavily on the results of research for their prosperity and progress.

Consider the quite parallel case of the manufacturing industries—such as the chemical industry represented, say, by Du Pont; the oil industry represented by Exxon; and the photography industry represented by Eastman Kodak. Each of these companies has been outstandingly successful for a long time, and one of the primary reasons for this has been the introduction of new products that originated in laboratory research. Without research Du Pont would have no nylon, Exxon would have no catalytic cracking process, and Eastman would have no Kodachrome.

But laboratory experiments on a test-tube scale do not get translated into rolls of Kodachrome available everywhere until the policymaking executives of the organization decide that it would probably be a paying proposition. A key figure on the route from test tube to sales point is the director of industrial research. He is pivotal in translating information. He knows how the laboratory scientist thinks, and he knows how the operating executive thinks, and he also knows that the thinking of the two do not often mesh. He is therefore invaluable in representing the industry's business needs to the scientist, and in showing the financial man how research can turn a profit.

So far as I have been able to learn, no such liaison function exists between ocean scientists and ocean industry. It is quite true that much of the research information that is generated in university and government laboratories does find its way into practical application, but it would occur much more rapidly and efficiently if the role of the research director were recognized and implemented.

Not to belabor the point, the university-based scientist for the most part is immersed in his teaching and research, and he does not often speak the language of the man who most needs his help. The captain of the sardine purse-seiner that I sailed with out of San Francisco years ago spoke a fine Italian-flavored broken English. To imagine a dialogue between him and some university professors boggles the mind.

Of course, no single boat owner could afford a research director, but an association of them could, especially if the processors also joined. Thus it is interesting to speculate on what might have happened to the California sardine fishery if such an association had called the attention of such a research director to the seriousness of the decline. Perhaps the fishery could have been saved.

The promise of this approach, especially for problems in the social sciences, has been cogently argued by Dr. Harold Gershinowitz, who for many years was the senior research executive of the Royal Dutch Shell Group. He was the first chairman of the Environmental Studies Board of the National Academy of Science–National Academy of Engineering, and has served as a member of several panels that prepared advisory reports for the federal government. In *Science* for April 28, 1972, he contributed the thought-provoking article "Applied Research for the Public Good—A Suggestion." In summarizing his suggestion, he said:

"The techniques that have been developed for the application of physical science to technology have been outstandingly successful. It seems worth applying them or their analogs to both physical and social sciences in order to benefit the public. To do this, it will be necessary to bring together politicians, administrators, and research workers in a manner that encourages their interaction and communication. There is no magic formula for accomplishing this. The methods that have been successful are as diverse as the corporations or mission-oriented agencies in which they have been used. The successful methods will probably be as diverse as the governments and other sociopolitical entities that make use of them. The common element is the recognition that the application of research is a complex operation, involving continuing interaction and feedback, and is not a simple orderly process of transmitting information from one place to another."

THE NATIONAL SCIENCE BOARD. The National Science Board is required by public law to report annually to the President of the United States "an appraisal of the status and health of science, as well as that of the related matters of manpower and other resources, in reports to be forwarded to the Congress."

The Board is composed of twenty-five members of note in both the academic and the industrial worlds: research professors, university presidents, and corporate leaders. In 1971 the chairman was Dr. H. E. Carter, Vice-Chancellor for Academic Affairs at the University of Illinois, and the Board's report in 1971 was entitled *Environmental Science: Challenge for the Seventies*. In its letter of transmittal to the President, the Board said:

> In choosing environmental science as the topic of this Report, the National Science Board hopes to focus attention on a critical aspect of environmental concern, one that is frequently taken for granted, whose status is popularly considered to be equivalent to that of science generally, and yet one whose contribution to human welfare will assume rapidly growing importance during the decades immediately ahead.
>
> The National Science Board strongly supports the many recent efforts of the Executive Branch, the Congress, and other public and private organizations to deal with the bewildering array of environmental problems that confront us all. Many of these problems can be reduced in severity through the use of today's science and technology by an enlightened citizenry. This is especially true of many forms of pollution and environmental degradation resulting from overt acts of man. Ultimate solutions to these problems, however, will require decisive steps forward in our scientific un-

derstanding and predictive skills, and in our ability to develop the wisest control and management technologies.

There is in addition a much larger class of environmental phenomena with enormous impact, today and in the future, on man's personal and economic well-being. These phenomena extend from fisheries to forests. They include the natural disasters of hurricanes and tornadoes; earthquakes and volcanoes; floods, drought, and erosion. They encompass problems in the conservation of our resources of water, minerals, and wildlife. Included too are the more subtle effects of civilization on weather and climate, as well as many forms of natural pollution, such as allergens, environmental pests and diseases, and volcanic dust. Together, these phenomena share a common characteristic: they can be fully understood, predicted, and modified or controlled only by studying them in terms of the complex environmental systems of which they are a part. Such studies, however, have become possible only in recent years. Greatly expanded efforts will be required to understand the forces involved in the confrontation between man and his natural environment.

This Report is presented as a contribution to the decisions that need to be made if environmental science is to become a fully effective partner in society's effort to ensure a viable world for the future.

In its assignment of reporting on the "status and health of science," the National Science Board could give only a part of its attention to our central interest in the ocean environment, but a following report, *Patterns and Perspectives in Environmental Science,* based on contributions of many authorities, deals in detail with all of the areas that the Board's letter of transmittal directed attention to. It is encouraging to note that the contribution on the marine environment was made by Joel W. Hedgpeth.

THE CONCERNED CITIZEN AGAIN. In closing, we return to the role of the concerned citizen, who might be represented by the concerned citizen who opened this chapter, or by the teacher, or by the research director, or by a member of the National Science Board, or by a member of the Executive Branch or of Congress.

Each in his own way, and in whatever ecological niche he may find himself, can make a significant contribution to the preservation and enhancement of our ocean environment. He can inform himself conscientiously, he can take the time and trouble to form his own judgment on critical issues, and he can communicate his views to those who need and value his opinions—whether it be his next-door neighbor or his representative in Congress.

Thus he will be making his own unique contribution to cooperation between man and man, and between man and nature, which, if this book has made its point, is essential to human happiness and human progress.

He will have his own share in "the effecting of all things possible."

SOURCES

The sources that I found most helpful in preparing this book are listed below. A few of them are rather technical, but most are well within the range of the intelligent general reader. Some are outstanding in interest and value, and these I have mentioned in the text. The list makes no pretense of being a review of marine science, but the bibliographies that the major works contain will open up the entire field for anyone.

CHAPTER 1

ARNON, D. I.: "Discussion," in A. A. Buzzati-Traverso, ed.: *Perspectives in Marine Biology* (Berkeley: University of California Press; 1958), p. 24.

DE BISSCHOP, ERIC: *Tahiti Nui* (New York: McDowell, Obolensky; 1959).

EBERT, JAMES D.: "The Effecting of All Things Possible," *American Zoologist*, Vol. XII (February 1971).

HEDGPETH, JOEL W.: *Seashore Life of the San Francisco Bay Region and the Coast of Northern California* (Berkeley: University of California Press; 1962, 1970).

KUHN, THOMAS S.: *The Structure of Scientific Revolutions* (Chicago: University of Chicago Press; 2nd ed., 1970).

MCCONNAUGHEY, BAYARD H.: *Introduction to Marine Biology* (St. Louis: The C. V. Mosby Company; 1970).

MOORE, HILARY B.: *Marine Ecology* (New York: John Wiley & Sons, Inc.; 1958).

MORISON, SAMUEL ELIOT: *The European Discovery of America; The Northern Voyages,* A.D. *500–1600* (New York: Oxford University Press; 1971).

SAUER, CARL O.: "Seashore—Primitive Home of Man?" in John Leighly, ed.: *Land and Life: A Selection from the Writings of Carl Ortwin Sauer* (Berkeley: University of California Press; 1963).

SHACKLETON, EDWARD: "Men Against the Sea," in G. E. R. Deacon, ed.: *Seas, Maps, and Men: An Atlas History of Man's Exploration of the Oceans* (Garden City, N.Y.: Doubleday & Company, Inc.; 1962).

STEPHENS, WILLIAM M.: *Science Beneath the Sea* (New York: G. P. Putnam's Sons; 1966).

VILLEE, CLAUDE A.: *Biology* (Philadelphia: W. B. Saunders Company; 1967).

WEYL, PETER K.: *Oceanography: An Introduction to the Marine Environment* (New York: John Wiley & Sons, Inc.; 1970).

CHAPTER 2

ALEXANDER, MARY CHARLOTTE: *The Story of Hawaii* (New York: American Book Company; 1912).

BECKWITH, MARTHA WARREN: *The Kumulipo: A Hawaiian Creation Chant* (Chicago: University of Chicago Press; 1951).

BUCK, PETER H.: *Vikings of the Sunrise* (New York: Frederick A. Stokes Company; 1938).

CHICKERING, WILLIAM H.: *Within the Sound of These Waves: The Story of Old Hawaii* (New York: Harcourt, Brace and Company; 1941).

DAY, A. GROVE, and CARL STROVEN: *A Hawaiian Reader* (New York: Appleton-Century-Crofts, Inc.; 1959).

DE BISSHOP, ERIC: *Tahiti Nui* (New York: McDowell, Obolensky; 1959).

EMERSON, NATHANIEL B.: *Unwritten Literature of Hawaii* (Washington, D.C.: U.S. Government Printing Office; 1909).

FURNAS, J. C.: *Anatomy of Paradise: Hawaii and the Islands of the South Seas* (New York: William Sloane Associates, Inc.; 1948).

GATTY, HAROLD: *Nature Is Your Guide: How to Find Your Way on Land and Sea by Observing Nature* (New York: E. P. Dutton & Company, Inc.; 1958).

JUDD, HENRY P., MARY KAWENA PUKUI, and JOHN F. G. STOKES: *Introduction to the Hawaiian Language* (Honolulu: Tongg Publishing Company; 1945).

KUYKENDALL, RALPH S., and A. GROVE DAY: *Hawaii: A History* (Englewood Cliffs, N.J.: Prentice-Hall, Inc.; 1948, 1961).

WARD, RITCHIE R.: *The Living Clocks* (New York: Alfred A. Knopf, Inc.; 1971).

CHAPTER 3

BROWN, LLOYD A.: *The Story of Maps* (Boston: Little, Brown and Company; 1949).

DALE, PAUL W.: *Seventy North to Fifty South: The Story of Captain Cook's Last Voyage* (Englewood Cliffs, N.J.: Prentice-Hall, Inc.; 1969).

HOLMES, SIR MAURICE: "Captain James Cook, R.N., F.R.S.," *Endeavour*, Vol. VIII (January 1949).

KUYKENDALL, RALPH S., and A. GROVE DAY: *Hawaii: A History* (Englewood Cliffs, N.J.: Prentice-Hall, Inc.; rev. ed., 1961).

PRICE, A. GRENFELL: *The Explorations of Captain James Cook in the Pacific, as told by Selections of his own Journals, 1768–1779* (New York: The Heritage Press; 1958).

SAMWELL, DAVID: *A Narrative of the Death of Captain James Cook* (London: G. C. J. and J. Robinson; MDCCLXXXVI. Hawaiian Historical Society Reprints No. 2, 1779).

SKELTON, R. A.: *Captain James Cook—after two hundred years* (London: The Trustees of the British Museum; 1969).

WALLBANK, T. WALTER, ALASTAIR M. TAYLOR, and GEORGE BARR CARSON, JR.: *Civilization, Past and Present*, Vol. II (Glenview, Illinois; 5th ed., 1965).

CHAPTER 4

ARNON, D. I.: "Discussion," in A. A. Buzzati-Traverso, ed.: *Perspectives in Marine Biology* (Berkeley: University of California Press; 1958), pp. 23–4. Quotation used by permission of the Regents of the University of California.

ASIMOV, ISAAC: *The Intelligent Man's Guide to Science* (New York: Basic Books, Inc.; 1972).

BUZZATI-TRAVERSO, A. A.: *Perspectives in Marine Biology* (Berkeley: University of California Press; 1958).

GABRIEL, MORDECAI L., and SEYMOUR FOGEL: *Great Experiments in Biology* (Englewood Cliffs, N.J.: Prentice-Hall, Inc.; 1955).

GIBBS, F. W.: *Joseph Priestley; Revolutions of the Eighteenth Century* (Garden City, New York: Doubleday & Company, Inc.; 1967).

INGENHOUSZ, JAN: *An Essay on the Food of Plants and the Renovation of Soils* (reprinted for limited distribution by J. Christian Bay, Oquawka, Illinois; 1933).

KING-HELE, DESMOND GEORGE: "Joseph Priestley," in Trevor I. Williams, ed.: *A Biographical Dictionary of Scientists* (London: A. and C. Black, Ltd; 1969).

MORRIS, CONSTANCE LILY: *Maria Theresa, the Last Conservative* (New York: Alfred A. Knopf, Inc.; 1937).

PRIESTLEY, JOSEPH: "Observations on Different Kinds of Aire," *Philosophical Transactions of the Royal Society of London*. Vol. LXII (1772).

———: *Memoirs of Dr. Joseph Priestley, Written by Himself* (London: H. R. Allenson; 1904).

RAPHAEL, SANDRA JOAN: "Jan Ingenhousz," in Trevor I. Williams, ed.: *A Biographical Dictionary of Scientists* (London: A. and C. Black, Ltd; 1969).

REDFIELD, ALFRED C.: "The Inadequacy of Experiment in Marine Biology," in A. A. Buzzati-Traverso, ed.: *Perspectives in Marine Biology* (Berkeley: University of California Press; 1958), pp. 17–23. Used by permission of the Regents of the University of California.

REED, HOWARD S.: "Jan Ingenhousz, Plant Physiologist. With a History of the Discovery of Photosynthesis," *Chronica Botanica*, Vol. XI (1949).

SCHOFIELD, ROBERT E.: *A Scientific Autobiography of Joseph Priestley* (Cambridge, Mass.: M.I.T. Press; 1966).

WIESNER, JULIUS: *Jan Ingen-Housz; Sein Leben und Sein Wirken als Naturforscher und Arzt* (Vienna: Carl Konegan; 1905).

CHAPTER 5

ELTON, CHARLES: *The Ecology of Animals* (London: Methuen & Co., Ltd; 1933). This excellent little book has survived three editions and many reprintings, and is now available in the "Science Paperback" series, distributed in the United States by Barnes & Noble, Inc. Readers who may have been annoyed and confused by hearing the term "ecology" applied to everything from the collection of empty beer cans to the design of nuclear power plants will find it excellent reading. It can be scanned in an evening.

FORBES, EDWARD: *History of British Starfishes* (London: John van Voorst; 1841).

———: "Report on the Mollusca and Radiata of the Aegean Sea, and on Their Distribution, Considered as a Bearing on Geology," *Report of the 13th Meeting, British Association for the Advancement of Science, 1843* (1844).

———: *The Natural History of the European Seas* (London: John van Voorst; 1859). This volume was published posthumously. Forbes himself wrote the first 102 pages as they appeared in print, and Robert Goodwin-Austen completed the work according to a plan that Forbes had left in outline.

HAECKEL, ERNST: *"Über Entwickelungsgang und Aufgabe der Zoologie" Jenaische Zeitung*, Vol. V. Quoted in W. C. Allee, Alfred E. Emerson, Orlando Park, Thomas Park, and Karl P. Schmidt: *Principles of Animal Ecology* (Philadelphia: W. B. Saunders Company; 1949).

HEDGPETH, JOEL W.: "A Century at the Seashore," *Scientific Monthly*, Vol. LXI (1945).

———: "Introduction," in *Treatise on Marine Ecology and Paleoecology*, Vol. I (New York: The Geological Society of America; 1957).

HERDMAN, SIR WILLIAM A.: *Founders of Oceanography and Their Work* (London: Edward Arnold & Co.; 1923).

RITCHIE, J.: "A Double Centenary," *Proceedings of the Royal Society of Edinburgh*, Section B, Vol. LXVI (1956).

THOMSON, SIR CHARLES WYVILLE: *The Depths of the Sea* (London: Macmillan and Co.; 1874).

WILSON, GEORGE, and ARCHIBALD GEIKIE: *Memoir of Edward Forbes, F.R.S.* (Cambridge, London, and Edinburgh: Macmillan & Co., Ltd; 1861).

CHAPTER 6

AGASSIZ, ELIZABETH CARY: *Louis Agassiz: His Life and Correspondence*, Vols. I and II (Boston: Houghton Mifflin Company; 1886).

COOPER, LANE: *Louis Agassiz as a Teacher* (Ithaca, N.Y.: Comstock Publishing Company, Inc.; 1945).

JAMES, WILLIAM: "Louis Agassiz," *Science*, N.S. Vol. V (1897).

JORDAN, DAVID STARR: "Agassiz at Penikese," *Popular Science Monthly*, Vol. XL (1892).

LOWELL, JAMES RUSSELL: "Agassiz," *Atlantic Monthly*, Vol. XXXIII (1874).

LURIE, EDWARD: *Louis Agassiz: A Life in Science* (Chicago: University of Chicago Press; 1960).

MARCOU, JULES: *Life, Letters, and Works of Louis Agassiz*, Vol. I and II (New York: Macmillan and Co., 1896).

MAYR, ERNST: "Agassiz, Darwin, and Evolution," *Harvard Library Bulletin*, Vol. XIII (Spring, 1959).

SCUDDER, SAMUEL H.: "In the Laboratory with Agassiz," *Every Saturday*, Vol. XVI (April 4, 1874).

CHAPTER 7

ASIMOV, ISAAC: "In Dancing Flames a Greek Saw the Basis of the Universe," *Smithsonian*, Vol. II (November 1971).
————: *Asimov's Biographical Encyclopedia of Science and Technology* (Garden City, N.Y.: Doubleday & Company, Inc.; 1964).
HARRISON, GEORGE RUSSELL: *The Conquest of Energy* (New York: William Morrow & Company; 1968).
KNIGHT, DAVID MARCUS: "Julius Robert von Mayer," in Trevor I. Williams, ed.: *A Biographical Dictionary of Scientists* (London: A. and C. Black, Ltd; 1969).
KUHN, THOMAS S.: "Conservation of Energy as an Example of Simultaneous Discovery," in Marshall Clagett, ed.: *Critical Problems in the History of Science* (Madison: University of Wisconsin Press; 1959).
————: *The Structure of Scientific Revolutions* (Chicago: University of Chicago Press; 2nd ed., 1970).
TATON, RENÉ: *History of Science: Science in the Nineteenth Century* (New York: Basic Books, Inc.; 1965).
TYNDALL, SIR JOHN: *Fragments of Science* Vol. I (New York: D. Appleton and Company; 1899).
VILLEE, CLAUDE A.: *Biology* (Philadelphia: W. B. Saunders Company; 1967).

CHAPTER 8

DEACON, MARGARET: *Scientists and the Sea* (New York: Academic Press; 1971).
HARDY, SIR ALISTER: *The Open Sea: Its Natural History* (Boston: Houghton Mifflin Company; 1965), Part I: *The World of Plankton*.
HERDMAN, SIR WILLIAM A.: *Founders of Oceanography and Their Work* (London: Edward Arnold & Co.; 1923).
MARSHALL, NORMAN B.: *Ocean Life in Color* (New York: The Macmillan Company; 1971).
MCCONNAUGHY, BAYARD H.: *Introduction to Marine Biology* (St. Louis: The C. V. Mosby Company; 1970).
MOSELEY, H. N.: *Notes by a Naturalist on the "Challenger"* (London: Macmillan and Co.; 1879).
RAPHAEL, SANDRA JOAN: "Sir Charles Wyville Thomson," in Trevor I. Williams, ed.: *A Biographical Dictionary of Scientists* (London: A. and C. Black, Ltd; 1969).

THOMSON, SIR CHARLES WYVILLE: "On the Embryology of *Antedon rosaceus*, Linck (*Comatula rosacea* of Lamarck)," *Philosophical Transactions of the Royal Society of London,* Vol. CLV (1865).
———: *The Depths of the Sea* (London: Macmillan and Co.; 1874).
———: *The Voyage of the "Challenger": The Atlantic,* Vols. I and II (London: Macmillan and Co., 1877).

CHAPTER 9

HERDMAN, SIR WILLIAM A.: *Founders of Oceanography and Their Work* (London: Edward Arnold & Co.; 1923).

CHAPTER 10

BAKER, FRED: "Dr. Ritter and the Founding of the Scripps Institution of Oceanography," Bulletin No. 15 (Non-Technical) of the Scripps Institution of Oceanography of the University of California; January 18, 1928.
GOSSE, PHILIP HENRY: *A Naturalist's Rambles on the Devonshire Coast* (London: John van Voorst; 1853).
HEDGPETH, JOEL W.: "*De Mirabili Maris:* Thoughts on the Flowering of Seashore Books," in *Proceedings of the Second International Congress on the History of Oceanography. Proceedings of the Royal Society of Edinburgh* B, Vol. LXXII (1971/72).
HERDMAN, SIR WILLIAM A.: *Founders of Oceanography and Their Work* (London: Edward Arnold & Co.; 1923).
KOFOID, CHARLES ATWOOD: *The Biological Stations of Europe* (Washington, D.C.: U.S. Government Printing Office; 1910).
LILLIE, FRANK R.: *The Woods Hole Marine Biological Laboratory* (Chicago: University of Chicago Press; 1944).
RAITT, HELEN, and BEATRICE MOULTON: *Scripps Institution of Oceanography: First Fifty Years* (Los Angeles: The Ward Ritchie Press; 1967).
WHITMAN, CHARLES O.: "Some of the Functions and Features of a Biological Station," *Science,* Vol. VII (1898).
YONGE, C. M.: "Development of Marine Biological Laboratories," *Science Progress,* Vol. XLIV (1956).

CHAPTER 11

ADAMS, HENRY: *The Education of Henry Adams* (New York: The Modern Library; 1931).
AGASSIZ, ELIZABETH C., and ALEXANDER AGASSIZ: *Seaside Studies in Natural History* (Boston: James R. Goodwin and Company; 1871).

AGASSIZ, G. R.: *Letters and Recollections of Alexander Agassiz, with a Sketch of His Life and Work* (Boston: Houghton Mifflin Company; 1913).
BATES, MARSTON: *The Forest and the Sea: A Look at the Economy of Nature and the Ecology of Man* (New York: Random House, Inc.; 1960).
DARWIN, CHARLES: *The Structure and Distribution of Coral Reefs* (New York: D. Appleton and Co.; 1842).
HERDMAN, SIR WILLIAM A.: *Founders of Oceanography and Their Work* (London: Edward Arnold & Co.; 1923).
MCCONNAUGHY, BAYARD H.: *Introduction to Marine Biology* (St. Louis: The C. V. Mosby Company; 1970).
RAITT, HELEN, and BEATRICE MOULTON: *Scripps Institution of Oceanography: First Fifty Years* (Los Angeles: The Ward Ritchie Press; 1967).
WEYL, PETER K.: *Oceanography: An Introduction to the Marine Environment* (New York: John Wiley & Sons, Inc.; 1970).
YONGE, C. M.: "Ecology and Physiology of Reef-Building Corals," in A. A. Buzzati-Traverso, ed.: *Perspectives in Marine Biology* (Berkeley: University of California Press; 1958).

CHAPTER 12

HERDMAN, SIR WILLIAM A.: "Oceanography, Bionomics, and Aquiculture," *Report of the British Association for the Advancement of Science,* Vol. LII, No. 1351 (1895).

CHAPTER 13

BURTON, MAURICE: "Life in the Sea," in G. E. R. Deacon, ed.: *Seas, Maps, and Men: An Atlas History of Man's Exploration of the Oceans* (Garden City, N.Y.: Doubleday & Company, Inc.; 1962).
HARDY, SIR ALISTER: *The Open Sea: Its Natural History* (Boston: Houghton Mifflin Company; 1956).
HERDMAN, SIR WILLIAM A.: *Founders of Oceanography and Their Work* (London: Edward Arnold & Co.; 1923).
HJORT, JOHAN: "Fluctuations in the Great Fisheries of Northern Europe," *Rappt. et Proc. verb. conseil permanent intern. exploration mer,* Vol. XX (1914).
———: *The Human Value of Biology* (Cambridge: Harvard University Press; 1938).
ISAACS, JOHN D.: "The Nature of Oceanic Life," in *Oceanography: Readings from Scientific American* (San Francisco: W. H. Freeman and Company; 1971).

ORR, A. P., and S. M. MARSHALL: *The Fertile Sea* (London: Fishing News [Books] Limited; 1969).

ROBERTS, KENNETH: *I Wanted to Write* (Garden City, N.Y.: Doubleday & Company, Inc.; 1949).

SCHLEE, SUSAN: *The Edge of an Unfamiliar World: A History of Oceanography* (New York: E. P. Dutton & Co., Inc.; 1973).

UNDERWOOD, EDGAR ASHWORTH: "Viktor Hensen," in Trevor I. Williams, ed.: *A Biographical Dictionary of Scientists* (London: A. and C. Black, Ltd; 1969).

CHAPTER 14

ALLEE, W. C.: *Animal Aggregations: A Study in General Sociology* (Chicago: University of Chicago Press; 1931).

———: *The Social Life of Animals* (New York: W. W. Norton & Company, Inc.; 1938).

ASTRO, RICHARD: *From Cosmos to Chaos: Edward F. Ricketts and the Fiction of John Steinbeck* (Minneapolis: University of Minnesota Press; 1973).

BALDWIN, JAMES MACK, ed.: "Cause and Effect," in *Dictionary of Philosophy and Psychology,* Vol. I (Gloucester, Mass.: Peter Smith; 1957).

CABRERA, ANGEL: "Ecological Incompatibility: An Interesting Biological Law," *An. Soc. Cient. Argentina,* Vol. CXIV (1932). Abstracted in *Biological Abstracts,* No. 4488, March 1935.

CALVIN, JACK, and JOEL W. HEDGPETH: "Preface: About this Book and Ed Ricketts," in Edward F. Ricketts and Jack Calvin: *Between Pacific Tides* (Stanford, Calif.: Stanford University Press; 3rd ed., revised by Joel W. Hedgpeth, 1962).

ELTON, CHARLES: *The Ecology of Animals* (London: Methuen & Co., Ltd; 1933).

FONTENROSE, JOSEPH: *John Steinbeck: An Introduction and Interpretation* (New York: Barnes & Noble, Inc.; 1963).

HEDGPETH, JOEL W., ed.: *Treatise on Marine Ecology and Paleoecology* (Washington, D.C.: The Geological Society of America; 1957).

———: *Seashore Life* (Berkeley: University of California Press; 1962, 1970).

———: "Philosophy on Cannery Row," in Richard Astro and Tetsumaro Hayashi, eds.: *Steinbeck: The Man and His Work* (Corvallis: Oregon State University Press; 1971).

RAITT, HELEN, and BEATRICE MOULTON: *Scripps Institution of Oceanography: First Fifty Years* (Los Angeles: The Ward Ritchie Press; 1967).

RICKETTS, EDWARD F., and JACK CALVIN: *Between Pacific Tides* (Stanford, Calif.: Stanford University Press; 1939).

———: *Between Pacific Tides* (Stanford, Calif.: Stanford University Press; 4th ed., revised by Joel W. Hedgpeth, 1968).
STEINBECK, JOHN: *Cannery Row* (New York: The Viking Press; 1945).
———: *The Log from the Sea of Cortez* (New York: The Viking Press; 1951).
———, and EDWARD F. RICKETTS: *Sea of Cortez* (New York: The Viking Press; 1941).

CHAPTER 15

BARDACH, JOHN: *Harvest of the Sea* (New York: Harper & Row; 1968).
BURTON, MAURICE: "The Discovery of Plankton," in G. E. R. Deacon: *Seas, Maps, and Men: An Atlas History of Man's Exploration of the Oceans* (Garden City, N.Y.; Doubleday & Company, Inc.; 1962).
FRASER, JAMES: *Nature Adrift: The Story of Marine Plankton* (London: G. T. Foulis & Co., Ltd; 1962).
GUNN, NEIL M.: *The Silver Darlings* (New York: George W. Stewart, Publishers, Inc.; 1945).
HARDY, SIR ALISTER C.: "Toward Prediction in the Sea," in A. A. Buzzati-Traverso, ed.: *Perspectives in Marine Biology* (Berkeley: University of California Press; 1958).
———: *The Open Sea: Its Natural History* (Boston: Houghton Mifflin Company; 1965).
———: *Great Waters: A Voyage of Natural History to Study Whales, Plankton, and the Waters of the Southern Ocean* (New York: Harper & Row; 1967).
HEDGPETH, JOEL W., ed.: *Treatise on Marine Ecology and Paleoecology* (Washington, D.C.: The Geological Society of America; 1957).
———: Personal communication; November 7, 1972.
HODGSON, W. C.: "The Great Herring Mystery," *The New Scientist*, November 29, 1956.
HOLT, S. J.: "The Food Resources of the Ocean," in *Oceanography: Readings from Scientific American* (San Francisco: W. H. Freeman and Company; 1971).
RICKETTS, EDWARD F., and JACK CALVIN: *Between Pacific Tides* (Stanford, Calif.: Stanford University Press; 4th ed., revised by Joel W. Hedgpeth, 1968).

CHAPTER 16

DE KRUIF, PAUL: *Microbe Hunters* (New York: Harcourt, Brace and Company; 1926).
MCCONNAUGHEY, BAYARD H.: *Introduction to Marine Biology* (St. Louis: The C. V. Mosby Company; 1970).

VILLEE, CLAUDE A.: *Biology* (Philadelphia: W. B. Saunders Company; 1967).
WAKSMAN, SELMAN A.: "The Role of Bacteria in the Cycle of Life in the Sea," *Scientific Monthly*, Vol. XXXVIII (1934).
WOOD, E. J. F.: *Marine Microbial Ecology* (New York: Reinhold Publishing Corp.; 1965).
ZOBELL, CLAUDE E.: *Marine Microbiology* (Waltham, Mass.: The Chronica Botanica Company; 1946).
———: "Introduction to Marine Microbiology," in *Contributions to Marine Microbiology* (New Zealand Department of Scientific and Industrial Research Information, Series No. 22; 1959).
———: "The Domain of the Marine Microbiologist," in Carl H. Oppenheimer, ed.: *Symposium on Marine Microbiology* (Springfield, Ill.: Charles C. Thomas, Publisher; 1963).

CHAPTER 17

ASIMOV, ISAAC: *Asimov's Guide to Science* (New York: Basic Books, Inc.; 1972).
———: *Photosynthesis* (New York: Basic Books, Inc.; 1968).
BASSHAM, J. A., and MELVIN CALVIN: *The Path of Carbon in Photosynthesis* (Englewood Cliffs, N.J.: Prentice-Hall, Inc.; 1957).
BASSHAM, J. A., and RICHARD C. JENSEN: "Photosynthesis of Carbon Compounds," in Anthony San Pietro, Frances A. Greer, and Thomas J. Army, eds.: *Harvesting the Sun: Photosynthesis in Plant Life* (New York: Academic Press; 1967).
CALVIN, MELVIN: *Chemical Evolution: Molecular Evolution Towards the Origin of Living Systems on the Earth and Elsewhere* (New York: Oxford University Press; 1969).
———: *Chemical Evolution* (Eugene: Oregon State System of Higher Education; 1961).
FARBER, EDUARD: *Nobel Prize Winners in Chemistry, 1901–1961* (New York: Abelard-Schuman; 1963).
ROSENBERG, JEROME L.: *Photosynthesis: The Basic Process of Food-Making in Green Plants* (New York: Holt, Rinehart and Winston, Inc.; 1965).
VILLEE, CLAUDE A.: *Biology* (Philadelphia: W. B. Saunders Company; 1967).

CHAPTER 18

BEARDSLEY, MONROE C.: *Thinking Straight* (Englewood Cliffs, N.J.: Prentice-Hall, Inc.; 3rd ed., 1966).
DE WIT, C. T.: "Photosynthesis in Relationship to Overpopulation," in Anthony San Pietro, Frances A. Greer, and Thomas J. Army, eds.:

Harvesting the Sun: Photosynthesis in Plant Life (New York: Academic Press; 1967).

EHRLICH, PAUL R.: *The Population Bomb* (New York: Sierra Club/ Ballantine; rev. ed., 1971).

MADDOX, JOHN: "The Doomsday Syndrome," *Saturday Review of the Society*, Vol. LV (October 1972).

———: *The Doomsday Syndrome* (New York: McGraw-Hill Book Company; 1972).

———: "The Doomsday Syndrome," *Reader's Digest*, Vol. CII (February 1973).

CHAPTER 19

ASIMOV, ISAAC: *Asimov's Guide to Science* (New York: Basic Books, Inc.; 1972).

BENSON, ANDREW A., RICHARD F. LEE, and JUDD C. NEVENZEL: "Major Marine Metabolic Energy Sources," in R. M. S. Smellie, ed.: *Current Trends in the Biochemistry of Lipids* (New York: Academic Press; 1972).

———: Personal communication; November 10, 1972.

FULLER, NELSON: Personal communication; November 10, 1972.

LEE, RICHARD F., JED HIROTA, and ARTHUR M. BARNETT: "Distribution and Importance of Wax Esters in Marine Copepods and Other Zooplankton," *Deep-Sea Research*, Vol. XVIII (1971).

———, Judd C. Nevenzel, G. A. Paffenhöfer, and A. A. Benson, "The Metabolism of Wax Esters and Other Lipids by the Marine Copepod, *Calanus helgolandicus*," *Journal of Lipid Research*, Vol. XI (1971).

HARDY, SIR ALISTER: *The Open Sea: Its Natural History* (Boston: Houghton Mifflin Company; 1965).

HASLER, ARTHUR D., and JAMES A. LARSEN: "The Homing Salmon," in *Oceanography: Readings from Scientific American* (San Francisco: W. H. Freeman and Company; 1971).

MARSHALL, NORMAN B.: *Ocean Life in Color* (New York: The Macmillan Company; 1971).

SCRIPPS INSTITUTION OF OCEANOGRAPHY: Press release; November 11, 1971.

———: Press release; May 13, 1971.

CHAPTER 20

ALLEE, W. C.: "Keynote Address," *Ecological Monographs*, Vol. IV (1934).

ANON.: "Abalone Farm Flourishing," *P. G. & E. Progress*, Vol. XLVIII (March 1971).

BARDACH, JOHN: *Harvest of the Sea* (New York: Harper & Row; 1968).
———: "Aquaculture," *Science,* Vol. CLXI (September 13, 1968).
BASSHAM, JAMES A., and RICHARD G. JENSEN: "Photosynthesis of Carbon Compounds," in Anthony San Pietro, Frances A. Greer, and Thomas J. Army, eds.: *Harvesting the Sun: Photosynthesis in Plant Life* (New York: Academic Press; 1967).
DEDERA, DON: "Solving the Mysteries of the Seas," *The Humble Way,* Vol. X (First Quarter, 1971).
GROSS, M. GRANT: *Oceanography: A View of the Earth* (Englewood Cliffs, N.J.: Prentice-Hall, Inc.; 1972).
HESTER, FRANK J.: "New Approaches to Fishery Research," in Sidney Shapiro, ed.: *Our Changing Fisheries* (Washington, D.C.: U.S. Government Printing Office; 1971).
HOLT, S. J.: "The Food Resources of the Ocean," *Scientific American,* Vol. CCXXI (September 1969).
IDYLL, C. P.: *The Sea Against Hunger* (New York: Thomas Y. Crowell Company; 1970).
JONES, THOMAS W.: *Hermes, or the Future of Chemistry* (New York: E. P. Dutton & Company; 1928).
LONGWELL, A. CROSBY, and S. S. STILES: "The Genetic System and Breeding Potential of the Commercial American Oyster," *Endeavour,* Vol. XXIX (May 1970).
LUTEN, D. B.: "The Guidance of Systems," Address before the Shell Development Research Club, March 22, 1972.
———: "On Dubious Decision," *Proceedings of the Symposium on Water Balance in North America, American Water Resources Association Proceedings,* Series No. 7 (1969).
MAUGH, THOMAS H., II: "Chemistry in 1972: Not a Leap, But an Inch," *Science,* Vol. CLXXIX (February 9, 1973).
MAYR, ERNST: *Populations, Species, and Evolution* (Cambridge, Mass.: Harvard University Press; 1970).
PARSHLEY, H. M.: "Review of Books," *New York Herald Tribune,* July 24, 1929.
SHAPIRO, SIDNEY, ed.: *Our Changing Fisheries* (Washington, D.C.: U.S. Government Printing Office; 1971).
WALFORD, LIONEL A.: *Living Resources of the Sea: Opportunities for Research and Expansion* (New York: The Ronald Press Company; 1958).
WALSH, JOHN J.: "Implications of a Systems Approach to Oceanography," *Science,* Vol. CLXXVI (June 2, 1972).
WATT, K. E. F., ed.: *Systems Analysis in Ecology* (New York: Academic Press; 1966).

CHAPTER 21

GERSHINOWITZ, HAROLD: "Applied Research for the Public Good—A Suggestion," *Science,* Vol. CLXXVI (April 28, 1972).

HEDGPETH, JOEL W.: "Preface," to Edward F. Ricketts and Jack Calvin, *Between Pacific Tides* (Stanford, Calif.: Stanford University Press; 4th ed., 1968).

———: *Seashore Life of the San Francisco Bay Region and the Coast of Northern California* (Berkeley: University of California Press; 1962, 1970).

KENYON, RICHARD L.: "The Royal Society," *Chemical and Engineering News* (August 8, 1960).

NATIONAL SCIENCE BOARD: *Environmental Science: Challenge for the Seventies* (Washington, D.C.: U.S. Government Printing Office; 1971).

INDEX

abalone, 297
Abbe, Ernst, 243
Adams, Henry: *The Education of Henry Adams*, 167
Aegean Sea, 67–8
Agassiz, Alexander, 8, 152–67, 292; background, 153; coral reefs, 161–5; as an engineer, 152, 157–8; expeditions of, 160–1, 162–5; starfish and sea urchins, 156, 162 n.; writings, 160
Agassiz, Mrs. Alexander, 155, 157, 160
Agassiz, Auguste, 74 n.
Agassiz, G. R., 158, 160, 164; *Letters and Recollections*, 156
Agassiz, Louis, 7, 71–4, 74 n., 75–8, 78 n., 79–84, 126, 141, 143, 144, 147, 152, 155, 156, 157, 160; background, 73–5; fossil fishes, 75–7; as a teacher, 71–3, 77–82, 83–4, 168; writings, 77
Agassiz, Mrs. Louis (Cécile Braun), 153
Agassiz, Mrs. Louis (Elizabeth Cary) (second wife), 160 and n.; writings, 160
Age of Discovery, 6–7, 12, 26

agriculture, 179, 280; pesticides, 258–61; photosynthesis, 252, 263–5, 281, 283
Aitken, Dr., 119
Aldrich, Admiral, 117
Allee, W. C., 199–200, 201, 298–9; *Animal Aggregations*, 200
Allman, Professor G. J., 101
American Chemical Society, 290–1
anchovies, 289
Anderson, John, 83
Anderson School of Natural History, *see* Penikese Island Laboratory
animals: abyssal, 66–7, 96, 118–19, 178; age, 186–7; aggregations, 199–200, 208; behavior, 7, 294–6; colorations and forms, 135, 176–8; desert, 278; diseases, 183, 239; *see also individual animals*
Anson, Commodore George, 29
Antarctica, 37–8, 124, 216–17
Antedon rosaceus, 97–8, 101
aquiculture, *see* fisheries
Arctic Ocean, 39, 124

i

Argyll, Duke of, 121
Aristotle, 45, 69, 99, 170, 206, 302
"Ark, The" (biological station), 126
Arnon, Dr. Daniel I., 45 and *n*., 244
Ascidians, 177–8, 204–5
Asimov, Isaac, 56
Astropecten tenuispinus, 106
Atlantic Ocean, 134–5; see also British seas
Australia, 30, 33, 36, 37
Austria, 139

Bacon, Sir Francis: *New Atlantis*, 10, 302
bacteriology, 179, 231–2; see also marine microbiology
Baffin Bay, 99
Baker, Bette, 272
Baker, Dr. Fred, 149–50
Balboa, Vasco Núñez de, 6
Baldwin, James Mark, 206 *n*.
Ball, Robert, 65
Banks, Sir Joseph, 31, 41
barnacles, 220, 223
Bassham, James A., 252, 281, 286
Bates, Henry Walter, 176
Bateson, William, 174
Beagle (ship), 161, 165
Beardsley, Monroe C.: *Thinking Straight*, 258–61, 262–3
Beebe, William, 199
Behring, Emil von, 231
Belfast, N. Ireland, 96, 97, 98, 101
Belgium, 139
Ben Nevis Observatory, Scotland, 126
Benson, Dr. Andrew A., 271–2, 275–6, 278
Bentham, Jeremy, 51
Bernard, Claude, 192
biochemistry, 7, 44, 252, 273
Biological Abstracts (magazine), 202, 203
biological stations, see seaside laboratories
bionomics, 172, 174–8, 184
Bisalputra, Thana, 272
Bisschop, Eric de, 5–6
Blackie, John, 126
blind goby, 147
Bonney, Thomas, 121, 132
Borlaug, Norman E., 283
Bottino, Dr. Nestor, 272
Boyce, Professor, 183–4
Boyle, Robert, 302
Brady, Professor H. B., 173
Brisinga, 106
British Association for the Advancement of Science, 62, 67, 68, 69, 130, 131–3, 139, 170, 171
British seas, 67, 69, 97, 105–6, 176, 178, 181–2, 184, 213

Brook, J. M., 99
Brooks, W. K.: *The Genus Salpa*, 173
Brown, Lloyd A., 30
Bruce, Maj.-Gen. Sir David, 118, 231
Buchan, Alexander, 126
Buchanan, J. Y., 109, 128; *Narrative of the Expedition*, 122
Buck, Sir Peter, 26; *Vikings of the Sunrise*, 14

Cabrera, Angel, 202, 203
Calanus, 217, 220, 221, 225, 267–8; *C. atlanticus*, 268; *C. finmarchicus*, 217, 268; *C. helgolandicus*, 268, 278
California, Gulf of, 205, 238, 292
California, University of, 147, 149; at Berkeley, 147, 245, 246
Calocaris macandreae, 178
Calver, Captain, 106
Calvin, Jack, 211; *Between Pacific Tides*, 150, 196, 197, 200–4, 205, 304
Calvin, Melvin, 253–5; photosynthesis, 244–6, 249–52, 271, 277–8, 281
Cambridge University, 139
Cannery Row (Monterey, Calif.), 195, 197, 199
carbon-14, 246, 249, 250–1, 278
Carnegie Foundation, 139
Carnegie Institution Research Laboratory, 164
Carpenter, Dr. William, 98, 107, 170; deep-sea exploration, 101–4, 116, 124
Carson, Rachel: *Silent Spring*, 257
Carter, Dr. H. E., 307
cartography, 30, 37
Cephalodiscus, 118
Ceylon, 165
Chadwick, James, 282
Chaetopod worms, 139
Challenger expedition, 7, 9, 69–70, 95–6, 108–12, 113, 114, 116–31 *passim*, 137, 162, 169, 170, 171, 172, 182, 193, 216, 220, 292, 300
Challengerida, 116
Chlorella, 250, 277
Christmas Island, 117–18
chromatography, 277–8; paper, 245, 246–9
chronometer, 7, 38
Chumley, James, 127
Clark University, 144
Clubb (scientist), 176
coccolithophores, 191
cod, 186 and *n*., 187, 190, 193, 220, 221
coelenterates, 97
Cohn, Ferdinand, 230
Colding, Ludwig A., 86
Columbia University, 139
Columbus, Christopher, 6, 26, 108, 170
Colville, Lord, 29
comb jellies, 223

INDEX · iii

conservation, 9, 68, 286–90, 308; energy, 85–94; misinformation about, 286; *see also* ecology
Conservation Foundation, 284; study of, 284–6
Cook, Captain James, 7, 27–42, 46, 108; background, 28–9; explorations, 31–41; Hawaii, 26, 39–40; as a scientist, 27, 30, 31–2, 37, 43 and *n*.
Cooper, Lane, 83
copepods, 172–3, 193, 204, 217, 223, 267, 269–78 *passim*, 291
coral, 101, 105, 161, 162, 164–5
coral reefs, 96, 116–17, 120–1, 123, 161–5
Cornell University, 83
Cousteau, Jacques, 256
crabs, 220, 223
Crane, Walter, 130
crinoids, 97, 101
Crosbie, Dr., 109
Crum-Brown, Professor, 115, 126
crustaceans, 66, 135, 173, 220, 223
Cuvier, Baron Georges, 50, 75, 77, 78 *n*.

Dade, Ernest, 215
Dance, Nathaniel, 28
Dannevig, Captain, 180
Darwin, Charles, 27, 42, 58, 78 *n*., 137–8, 158, 169, 170, 174, 176, 293; coral reefs, 117, 120, 121, 161–5; writings, 137, 138, 161, 172, 292
Daubney, Charles, 56
David, Edgeworth, 121
Davy, Sir Humphry, 51
Dedera, Don, 284
deep-sea dredging, 64–7, 68, 96, 98–112, 116–31 *passim*, 134–5, 170
del Cano, Juan Sebastián, 7
Dendronotus arborescena, 176–7
De Wit, Dr. C. T., 264–5, 280–1
diatoms, 190, 217, 220, 266–7, 269–77 *passim*, 291
Discovery expedition, 216, 217, 220, 226
Dittmar, William, 126
Dogger Bank, 213, 229
Dohrn, Anton, 137–41, 143, 144, 215
Dolphin (ship), 269, 272
Doppelmayr, Johann, 30

Easter Island, 6, 12, 26
Ebert, Dr. James D., 9–10
echini, 156, 162 *n*., 292
echinoderms, 105–6, 156
Echinus flemingii, 105
ecology, 7, 8, 57–8, 63–70, 184, 196, 204, 212; Gause's principle, 202; hysteria about, 227, 256–8, 261–3; incompatibility in, 202–3; pollution, 142–3, 227, 240, 256–7, 307, 308; population and distribution, 57, 64, 65, 66–8, 173–4, 186–91, 193, 194, 199–200, 208, 227–9, 267, 288; solving problems in,

227–9, 258–65; teaching, 304–5; *see also* conservation; environment
Edinburgh, University of, 60–1, 69, 101, 106, 114, 115, 116, 118, 169
Edwards, H. Milne, 153, 158
eel, fresh-water, 135
Ehrlich, Dr. Paul, 231, 263; *The Population Bomb*, 261
Einstein, Albert, 282
Eliot, Charles, 8, 152
Elton, Charles, 58, 64; *The Ecology of Animals*, 58, 196
Emerson, Ralph Waldo, 78
Endeavour (ship), 27, 31, 33–6
energy, 85–94; atomic, 282–3; conservation law, 7, 85–94; conversions, 86–9, 92–4; heat, 88–9, 92–4, 270–1; sources, 281–3; and sunlight, 45, 85, 89–90, 191, 193, 220, 223, 241; and work, 87–9
Enteromorpha, 173
environment, 63, 199, 201, 217, 288, 290–2, 307–8; adaption to, 174–8; carbon monoxide in, 290–1; citizens and, 303, 308–9; distribution, 125, 291–2; industry, 142, 307; misinformation about, 290–1; *see also* ecology
Eolis, 176–7
Ericson, Leif, 6
euphausians, 193, 268
Euphausia superba, 217
evolution, 4–5, 7, 137, 138, 139, 172, 174–8, 275, 292

Faeroe (Faroe) Channel ridge, 124–5
Faeroe Islands, 6, 102, 103, 106
fat and wax metabolism, 220, 267, 270, 274–8 *passim*
Fearnow, Ted, 257
Fermi, Enrico, 245, 283
Fielde, Adele M., 143
fish, *see* animals
fisheries, 68, 172, 217; decline, 186, 191, 194, 227–9, 287–90, 306; government and, 68, 180–3, 285, 287; overfishing, 180, 186, 194, 227, 281, 287–8; *see also* ecology
flagellates, 217, 220, 291
Fontenrose, Joseph: *John Steinbeck: An Introduction and Interpretation*, 206
food: chain, 220, 267–78 *passim*, 291; from the sea, 68, 178–84, 212–15, 265, 283–6, 291–2, 294–8; inorganic nutrients, 223, 239, 277, 291; for man, 252, 258, 261–5, 280–1, 283–6; protein, 281, 283, 285, 286, 296; web, 220–1
Forbes, Edward, 7, 57, 58–70, 106, 114, 128, 181, 213; background, 59–63; deep-sea belief, 68, 98, 99, 101; ecology, 57, 61–70; population dynamics,

Forbes, Edward (*cont.*)
  57, 64, 68; writings, 64; zones of, 66–8
Forel, F. A., 158
fossils, 69, 75–6, 97
Foundation for Ocean Research, 269
Fox, Denis, 273
France, 7, 29, 37, 179–80
Fraser, James: *Nature Adrift*, 229
Funafuti (atoll), 121
Furneaux, Captain Tobias, 37

Gabor, Dennis, 273 *n.*
Galigher, Albert E., 198
Galileo, 99
Galsworthy, John, 305
Garstang, Walter, 176
Gatty, Harold, 20
Gee (scientist), 234
Geikie, Archibald, 61, 115, 119, 126, 132, 136
genetics, 7, 44, 292, 297–8
Geographical Journal, 129
geology, 119–21, 122
George III, King of England, 30, 33, 38, 40, 52
Germany, 139
Gershinowitz, Dr. Harold, 306–7
Giard, Alfred, 176, 177
Globigerina, 99, 116, 118, 119
Goethe, Johann, 197
Goodwin, John, 62
Gosse, Philip Henry, 175; *A Naturalist's Rambles on the Devonshire Coast*, 150–1
Granton, Scotland, 126
Grassi, Giovanni, 231
Gray, Asa, 78
Great Britain, 7, 29–31, 32–3, 37–40, 139; and fisheries, 179–84
Greeks, 5, 12
Greenland, 6

Haeckel, Ernest, 57–8, 126, 129, 138, 158, 170, 218
Hakluyt Society, 27
Hardy, Sir Alister, 8, 212–17, 220–6, 232, 266; background, 213–15; food resources, 212; herring fisheries, 217, 224–7, 268; research instruments, 214, 224–7, 295; writings, 213–14, 216, 217, 224–5, 229
Harriman, E. H., 148
Harrison, John, 7, 38
Harvard University, 8, 78, 83; Museum of Comparative Zoology, 78, 83, 152, 153, 155, 157, 160, 166, 186
Harvesting the Sun: Photosynthesis in Plant Life (book), 263–4
Hasler, Arthur D., 274
Hastigerina, 116
Hawaii, 6, 12, 14, 24–6, 39–40, 164
Hawaii-loa, 14–15, 43; his migration to

Hawaii, 6, 14, 15–26
Hawthorne, Nathaniel, 78
heat as energy, 88–9, 92–4, 270–1
Hedgpeth, Joel W., 197, 202, 203 and *n.*, 211, 217, 304–5, 308; revised *Between Pacific Tides*, 150, 203, 204, 304; as a teacher, 304–5; writings, 196 *n.*, 200, 204, 304
Helgoland, 187
Helland-Hansen, Dr. Björn, 134
Helmholtz, Herman von, 86
Hensen, Viktor, 191–3; biomass, 192; plankton, 191–3, 218
Herdman, Sir William, 8, 69, 95, 96–7, 106, 113, 130, 164; aquiculture, 172, 178–84; bionomics, 172, 174–8, 184; *Founders of Oceanography and Their Work*, 113–36, 139, 161, 164–5, 192; oceanography, 169–74, 184
herrings, 186 and *n.*, 187, 189, 193, 212, 215; fisheries, 213–15, 217, 224–9, 295; food of, 172, 191, 217, 220–1, 268
Hester, Frank J.: *Our Changing Fisheries*, 295–6
Heyerdahl, Thor, 257
Higginson, H. L., 167
Hillier, James, 243
Hirota, Jed, 272
Hjort, Dr. Johan, 129, 134, 185–91, 193–4; fish population, 186–91, 194; writings, 125, 134, 186, 194
Hodgson, Dr. W. C., 229
Holland, 139, 179–80
Holmes, Oliver Wendell, 78
Holm-Hansen, Dr. Osmund, 272
holography, 273 and *n.*
Holopus, 118
Holothurians (Elasipoda), 118
Holt, S. J., 281, 291, 294
Holteniae, 106
Hopkins Marine Station, 196 and *n.*, 204, 205
Horne, Dr. John, 119
Hoyle, Dr. W. E., 128
Huahine (island), 37
Hungary, 139
Huxley, Thomas, 121, 158, 160
Hyalonema, 106
Hyde, Dr. Ida, 139

Iceland, 6, 12, 102
Ingenhousz, Arnold, 51
Ingenhousz, Jan, 46, 51–5, 89; background, 51–2; photosynthesis, 52–5, 270; smallpox inoculation, 51 and *n.*, 52; writings, 52, 53–5
Ingenhousz, Mrs. Jan, 52
International Congress for the Exploration of the Sea, 131
International Council for the Study of the Sea, 194

Irish Sea, 62, 182
Irvine, Robert, 126
Issatchenko (scientist), 234
Italy, 7, 139

Jackson, Sir John, 126
Jacquin, Nikolaus, 52
James, William, 83-4
Jardin des Plantes (museum), Paris, 154, 158-9
Java, 87
Jeffreys, Gwyn, 116, 181
jellyfish, 69, 223
Jenkin, Fleeming, 101
Jenner, Edward, 51 n.
Jensen, Richard G., 252, 281, 286
Johnson, Martin, 207
Johnson Oyster Company, 297
Jones, Thomas William: *Hermes, or the Future of Chemistry*, 282-3
Jordan, David Starr, 72, 84
Joule, James, 86, 92
Judd (scientist), 121

Kalaniopuu, 39
Kendall, Larcum, 38
Kingsley (scientist), 175
Kittredge, James, 273
Koch, Robert, 230
Kofoid, Charles A., 140, 166; *The Biological Stations of Europe*, 138, 142-3
krill, 193, 217
Kuhn, Thomas, 9, 10; *The Structure of Scientific Revolutions*, 90
Kuykendall, Ralph S., 39

Labrador, 29
lakes, 128-9
Lamellaria perspicua, 177
Larsen, James A., 274
Lasker, Reuben, 273
Lavoisier, Antoine, 50 and n., 53, 55, 87
Lawrence Radiation Laboratory, 245, 271
Lea, Einar, 187, 190 n.
Le Conte, Joseph, 147
Lee, Richard, 272
Leeuwenhoek, Anton van, 242
Leighton, Dr. David, 297
Leith, Emmet N., 273 n.
Leptoclinum maculatum, 177
Leuckart, Rudolf, 144
Lew Islands, 106
Liebig, Julius, 90
Lightning (ship), 103, 106, 107, 116, 124, 171
Lillie, Frank R., 143, 146
Lister, Joseph, 230, 242
Liverpool, University of, 113, 133, 170, 183
lochs: fresh-water, 128-9; sea-, 127-8

Loeb, Jacques, 143
Lofoten Islands, 101, 191
Longfellow, Henry Wadsworth, 78
Longwell, A. Crosby, 297
Lovejoy, Ritchie, 201
Lowell, James Russell, 78, 84
Lowell Institute, 77-8, 131, 132
Lurie, Edward: *Louis Agassiz: A Life in Science*, 77
Luten, Dr. D. B., 298

Maddox, John: *The Doomsday Syndrome*, 257
Magellan, Ferdinand, 6, 7, 26, 108, 170
Makalii (Hawaii-loa's navigator) 15, 19, 20, 22, 23
man, 253-5; evolution, 4-5; food for, 252, 258, 261-5, 280-1, 283-6; metabolism, 270, 275, 278-9; population, 186, 189-90, 227, 258, 261-5, 280-1, 283; thinking, creative and critical, 259-61
Maria Theresa, Empress of Austria, 52, 53
marine biology, 7, 43-5, 118, 141-3, 169; see also ecology; marine microbiology; oceanography; photosynthesis; seaside laboratories
marine microbiology, 44, 179, 232-40; marine bacteria, 232, 235-7, 238-40; see also marine biology
Martin, A. J. P., 246
Maskelyne, Nevil, 37
Masson, David, 126
Maugh, Thomas H., II, 291
May, Staff-Commander, 103
Mayer, Dr. A. G., 164
Mayer, Robert, 7, 86-94; background, 87; energy conversions, 86-9, 92-4, 270-1; writings, 92
Mayr, Ernst, 78 n.; *Populations, Species, and Evolution*, 293
McIntosh, Professor, 179, 181
McLaren, Lord, 126
Mediterranean Sea, 5, 11, 67, 138
medusa, 99
Medusa (ship), 127-8, 129
Metchnikoff, Elie, 231
Mexico, 205
Michael Sars (ship), 134-5, 186, 187
microbes, 230-1; *Escherichia coli*, 231; see also marine microbiology
microscopes, 232, 236-7, 238, 242-3; electron, 242, 243, 244
Mill, Dr. Hugh Robert, 126, 128
Millport biological station, 126, 129
mines, Calumet and Hecla, 152, 157-8
mollusks, 66, 69, 177, 223
Moore, Galen, 257
Moseley, H. N., 108, 109, 112, 173; *Narrative of the Expedition*, 122

Môtier, Switzerland, 73
Moulton, Beatrice, 206–7; *Scripps Institution of Oceanography*, 166
Mueller, Erwin W., 243
Mullen, Michael M., 277
Müller, Fritz, 158
Müller, Johannes, 156
Murray, Sir John, 7, 113–36, 158, 166, 167, 173, 182; background, 114–15; bipolarity, 124; *Challenger* reports, 111, 113, 114, 116, 117, 118, 119, 120, 121–31 *passim*, 169, 171, 172; coral reefs, 116–17, 120–1, 123, 162; deep-sea exploration, 109, 116–17, 119, 124–5, 126, 127–8, 130, 134–5; lochs, 127–9; oceanography, 113, 114, 122, 124, 125, 130, 134, 135; writings, 117, 122, 123, 134, 135
Muséum National d'Histoire Naturelle, Paris, 75

Nansen, Fridtjof, 129
Nares, Captain George S., 109
National Academy of Sciences, 272
National Research Council Committee, 304
National Science Board, 307–8
Nature (magazine), 121, 257
navigation, 5, 11–12; and bird migration, 22; celestial, 6, 12, 15, 19, 20, 31, 37; and trade winds, 20–2
nets: herring, 213–14; plankton, 192, 219–20, 224–7
Neuchâtel, Switzerland, 77, 153
Nevenzel, Dr. Judd C., 272
Newfoundland, 29
New Scientist, The (magazine), 229
Newton, Sir Isaac, 27, 42, 170
New Yorker, The (magazine), 257
New Zealand, 6, 12, 26, 30, 33, 36, 37
Nierenberg, Dr. William A., 284
Nineteenth Century (magazine), 121
Noel-Paton, Professor, 118
Norman, Canon, 129, 181
Norsemen, 6, 12
North America, 6, 39
Norway, 134, 179–80, 185–6
nuclear reactor, 245–6
Nudibranchiate Mollusca, 176

oceanography, 7–10, 69, 113, 122, 124, 129, 130, 133, 134, 135, 147, 168–74, 184; early studies, 7–8; and geology, 119–21, 122; research director in, 305–6; research instruments, 215–16, 224–7, 234, 235; sensing devices, 295–6, 299; systems analysis, 298–301; *see also* oceans and seas; seaside laboratories; *individual sciences*
oceans and seas, 184; abyss, 66–7, 96, 118–19, 173, 174, 178; beginning of life in, 137, 173; biomass, 192–3; coralline zone, 66, 178; deep-sea coral zone, 66–7, 178; deposits, 116, 117; ecology, 9–10; energy, 94; laminarian zone, 66, 173–4, 176, 178; light penetration, 98, 134–5, 192; littoral zone, 66, 174, 176, 178, 199, 202; mud-line, 134; pelagic life, 173; pressures, 98–9, 178, 240; salinity, 65–6, 239; universality of, 217; waves, 23; *see also* ecology; *individual oceans and seas*
Onslow, Earl of, 133
"oozes," 116, 118, 135
Orkney Islands, 106
Ottestad, Per, 187
Oxford University, 139
oysters, 297–8

Paao (Raitean high priest), 6, 24–6
Pacific Ocean, 6, 12, 19–23, 29–30, 31–3, 39, 147, 199, 202, 208
Paffenhöfer, Dr. Gustav A., 272, 277
Palliser, Sir Hugh, 29
Panthalis oerstedi, 178
Paris, France, 75; Academy of Sciences, 131
Parkinson (artist), 37
Parshley, H. M., 282
Pasteur, Louis, 230, 232
Peach, Dr. Benjamin, 119
Pearcey, F., 127
Pearson, Karl, 174
Penikese Island Laboratory, 71–2, 83, 141, 144
Pennsylvania, University of, 139
peridinians, 191
Perkins, John, 297
Peruvian anchovetas, 186 *n.*, 281, 300
pesticides, 258–61
Petersen, C. G. Joh., 130
Pettersson, Otto, 129
Petty, William, 302
Phoenicians, 5, 11
photometer, 134
photosynthesis, 7, 45, 46–50, 52–6, 192, 220, 238, 241–2, 249–52, 258, 262–5, 267, 270, 271, 278, 281, 283; chloroplasts, 242, 244, 249; tools for studying, 242–9
plaice, 187
plankton, 96, 116–17, 134, 135, 173, 191–3, 216, 217–27, 238, 239, 266, 291; *see also individual plankton*
plankton indicator, 224–5
plankton recorder, 216, 295; continuous, 225–7
plants, 217–24; *see also* plankton; *individual plants*
Plymouth Marine Biological Association, 182
Polynesians, 4, 5–6, 12–26

Porcupine (ship), 105–6, 107, 116, 124, 171
porpoises, 221
Port Erin Biological Station, 178, 183
prawns, 223
Prebus, Albert F., 243
Price, Dr. A. Grenfell, 28, 36–7
Priestley, Joseph, 7, 46–51, 89; background, 46; photosynthesis, 7, 47–50, 53, 270; writings, 46–7
Priestley, Timothy, 47
Pringle, Sir John, 51, 52
Pullar, Frederick P., 128
Pullar, Lawrence, 128
Pyrocystis, 116
Pyrosoma, 112 and *n.*, 119
Pytheas, 11–12

Rabinowitch, Eugene I., 56
Radiolaria, 116, 118
Raiatea (formerly Havaiki island), 6, 14, 15, 20, 26, 37
Raitt, Helen, 206–7; *Scripps Institution of Oceanography*, 166
Redfield, Dr. Alfred C., 44 and *n.*
Reed, Walter, 231
Renard, Abbé, 126, 127; *Deep-sea Deposits*, 117, 122
Richardson, I. D., 294–5
Ricketts, Edward F., 8, 195–211, 292, 298, 299; background, 197–8; *Between Pacific Tides*, 150, 196, 197, 200–4, 205, 304; ecology, 196, 199, 202–3, 208, 286; final and efficient causes, 206–8; *Sea of Cortez*, 205, 208; and John Steinbeck, 195–7, 204–11
Ritter, William Emerson, 147–50
Roberts, Kenneth: *I Wanted to Write*, 187 *n.*
Robertson, Dr. David, 129
Rosenberg, Dr. Abraham, 272
Ross, Sir John, 99
Ross, Sir Ronald, 231
Roumanian Academy of Sciences, 139
Roux, Pierre, 231
Royal Societies, 7, 28, 29, 31, 33, 37, 38, 41, 47, 90, 98, 101–5, 106, 107–8, 111, 116, 120, 121, 126, 181–2, 242, 302
Russell (scientist), 234
Russia, 139

Sabine, Gen. Sir Edward, 101–4
Sachs, Julius von, 56, 89
salmon, 269, 274; food of, 270, 274–8 *passim*
Samwell, Dr. David, 28, 40, 41–2
San Diego Marine Biological Association, 149
Sandwich Islands, 39
Sarcodictyon catenata, 178

sardines, 286–90
Sars, Michael, 101, 102
Sauer, Carl O., 4–5
Saussure, Horace de, 56, 89
Scandinavia, 7
Scarborough, Scotland, 213–15
Schmidt-Nielson, Dr. Knut, 278
science, 253–5, 281–3, 293; definitions, 4, 10; experimental method, 43–5, 169, 175, 206, 207–8, 302; naturalist approach, 43–5, 168–9, 175, 207, 302; revolutionary ideas in, 90; and technology, 253, 279, 308
Science (magazine), 291, 301, 306
Science Progress (magazine), 141
Scotland, 7, 120, 126; Fisheries Board, 180, 182; lochs of, 127–9
Scott (scientist), 173
Scottish Geographical Magazine, 129
Scottish Geographical Society, 126
Scottish Marine Biological Association, 226
Scottish Meteorological Society, 126
Scripps, E. W., 148, 166
Scripps, Ellen, 149
Scripps Institution of Oceanography, 44, 146, 147–50, 166, 206, 233, 269–78 *passim*, 284, 304
Scudder, Samuel H., 78–82, 84
sea butterflies, 223
seaside laboratories, 138–51, 174–8, 183, 184, 293; *see also* individual laboratories
sea urchins, 69, 156
Shaw, Quincy A., 157
Shelburne, Lord, 46, 51
shellfish, 67; *see also* individual shellfish
Shepard, Francis, 207
Shetland Islands, 62, 102, 106
shrimp, snapping, 207
Singer, Charles, 55, 220
Smith, Robert, 126
Smith, Theobald, 231
Smithsonian Institution, 139
snails, 69
Society Islands, 12, 26, 33
Solas, Professor, 121
SONAR, 207, 225
Spain, 29, 37
Spencer, Herbert, 175
sponges, 105–6, 178
squid, 223
Stanford University Press, 201–2, 203 *n.*, 204
Stanton, Hugh, 297
starfish, 64, 69, 156
Stazione Zoologica, 138–41, 147, 215
Steinbeck, John, 195–6, 196 *n.*, 197, 199, 204–5, 286, 292; final and efficient causes, 206–8; and Ed Ricketts, 195–

Steinbeck, John (*cont.*)
    7, 204–11; writings, 8, 195, 196, 197, 205, 208–11
Stevens, C. M., 291
Stevenson, Robert Louis, 126
Stornoway, Scotland, 105, 124
Strait of Georgia, 266, 267, 269, 277, 291, 298
Styles, S. S., 297
sun, 98, 123, 192; energy, 85, 89–90, 191, 193, 220, 223; photosynthesis, 7, 45, 53–6, 192, 220, 241, 263, 270
Swanson, Dr. Virginia, 272
Sweden, 102
Switzerland, 139
Synge, R. L. M., 246
syphonophores, 223
systematics, 293

Tahiti, 26, 31–2, 37, 39
Tait, P. G., 115, 116, 126
Tasman, Abel, 33
taxonomy, 292–3
Taylor, C. V., 204
Thompson (scientist), 173
Thompson, J. Vaughan, 219–20, 224
Thompson, William, 65
Thomson, Sir Charles Wyville, 7, 69–70, 95–112, 114, 119, 120, 121–2, 126, 131, 158, 162 *n.*, 169, 170–1, 181; background, 96; deep-sea exploration, 98, 101–12, 116, 124; Faeroe (Faroe) Channel ridge, 124–5; writings, 98, 104
Thomson, Isaac, 129
Thomson, Joseph, 119
Thomson, William, 92
Tigriopus fulvus, 204
Tisiphonia, 106
Tizard, T. H., 124–5; *Narrative of the Expedition*, 122
Tonga, 39

Tortugas, 164–5
Triton expedition, 125
Turner, Sir William, 115, 120, 126
Tyndall, Sir John, 90–2

United Nations Food and Agricultural Organization, 281, 283–4, 294
United States, 7, 139, 180, 262, 284
Upatnieks, Juris, 273 *n.*

Vienna, Austria, 52

Walford, Dr. Lionel A., 284–6, 289, 294–5, 296
Walker, Alfred O., 173
Walker, John and Henry, 28–9
Wallace, Alfred R., 176
Walsh, John J., 299–301
Ward, Ritchie: *The Living Clocks*, 19 *n.*
Waterson (scientist), 92
Watson, Arnold, 178
Weddell Sea, 216–17
Weldon, Professor, 174
West Indies, 6
whales, 172, 191, 193, 215, 216, 221, 278
Whitman, Dr. Charles Otis, 143–6
Wilke, Donald A., 272
Wilson (author), 61
Wilson (scientist), 234
Woods Hole, Mass., 199–200, 201
Woods Hole Marine Biological Laboratory, 9–10, 44, 83, 142–6, 147
Work and energy, 87–9
Wren, Christopher, 302

Yonge, Dr. C. M., 141

ZoBell, Dr. Claude E., 231–40; background, 232–3; J-Z sampler, 234; microbiology, 233–40; writings, 238

A NOTE ABOUT THE AUTHOR

Ritchie R. Ward was born in Medicine Lodge, Kansas, in 1906. He received his B.S. in chemistry and his M.Jour. from the University of California at Berkeley. After many years as a chemist and laboratory manager in Hawaii and California, he retired in 1962 and began a new career as a writer, instructor, and consultant on technical writing. Mr. Ward is the author of *Practical Technical Writing*, a textbook published in 1968, and has contributed biographical sketches of prominent scientists and engineers to *Current Biography*, published by H. W. Wilson. He also has prepared a writing seminar for scientists and engineers for closed-circuit TV, originating at Stanford University and available to sixteen local industrial facilities. His most recent book, *The Living Clocks*, was published in 1971. Mr. Ward has been a member of the U.S. National Committee for the International Federation for Documentation, a committee of the National Academy of Sciences, on which he served from 1961 to 1964. He is married, and lives in Orinda, California.

A NOTE ON THE TYPE

The text of this book was set on the Linotype in a type face called Baskerville. The face is a facsimile reproduction of types cast from molds made for John Baskerville (1706–75) from his designs. The punches for the revived Linotype Baskerville were cut under the supervision of the English printer George W. Jones. John Baskerville's original face was one of the forerunners of the type style known as "modern face" to printers—a "modern" of the period A.D. 1800.

This book was composed and bound at The Book Press, Brattleboro, Vt., and printed at Halliday Lithograph Corp., West Hanover, Mass. Typography and binding designed by The Etheredges.